Home Remote Control and Automation Projects

Second Edition

Home Remote Control and Automation Projects

Second Edition
Delton T. Horn

TAB Books
Division of McGraw-Hill, Inc.
New York San Francisco Washington, D.C. Auckland Bogotá
Caracas Lisbon London Madrid Mexico City Milan
Montreal New Delhi San Juan Singapore
Sydney Tokyo Toronto

© 1991 by **TAB Books**.
First edition © 1986 by TAB Books.
TAB Books is a division of McGraw-Hill, Inc.

pbk   5   6   7   8   9   10   11   12   13   14 FGR/FGR   9   9   8   7   6   5
hc   2   3   4   5   6   7   8   9   10   11 FGR/FGR   9   9   8   7   6   5   4   3   2

Library of Congress Cataloging-in-Publication Data

Horn, Delton T.
   Home remote-control and automation projects / by Delton T. Horn.
     p.   cm.
   Includes index.
   ISBN 0-8306-2197-0   ISBN 0-8306-2196-2 (pbk.)
    1. Dwellings—Automation. 2. Remote control. 3. Dwellings-
-Electric equipment.   I. Title.
   TH4812.H68   1991
   696—dc20   91-16747
                                            CIP

Acquisitions Editor: Roland S. Phelps
Managing Editor: Sandra Johnson Bottomley
Book Editor: Laura J. Bader
Director of Production: Katherine G. Brown
Book Design: Jaclyn J. Boone   EL2
Cover Design & Illustration: Greg Schooley, Mars, Pa.   3765

Contents

Projects

Preface

ELECTRONICS EXPERIMENTERS ENJOY PROJECTS FOR VARIOUS remote control and automation applications. Such projects are fascinating, educational, practical, and economical. They are also a great way to impress your friends and family with your hobby.

Besides being fun and educational, remote control and automation projects can often save you a great deal of money and effort. Let electronics do the job for you. For many handicapped persons, remote control and automation devices are a virtual necessity which can help them lead a more normal and satisfying life.

This new, expanded edition features over 70 practical remote control and automation projects. While each of the projects are fairly simple, they give impressive results. Most of these projects can be easily adapted in countless ways to suit other, specialized applications. Experimenting with these projects is strongly encouraged. You may find an application for one of these circuits that never even occurred to me.

The emphasis in this book is on the practical. While theoretical topics and general information on remote control and automation theory are briefly covered here, you may also want to read other books for more background information on this fascinating and exciting subject.

The basics of remote control and automation

THE PURPOSE OF THIS BOOK IS TO PRESENT A COLLECTION OF practical projects for remote control and automation applications. To fully understand the projects and allow you to customize them to your own individual applications, we'll start with two chapters on fundamental principles before getting into the actual projects. If you feel you already have a good background in this area, you may skip ahead to chapter 3.

Obviously, we can't go into too much detail in just two chapters. If you want to learn more about the underlying principles in this area, I recommend that you find *Handbook of Remote Control & Automation Techniques—2nd Edition* by John E. Cunningham and Delton T. Horn (TAB book 1777) in the library.

Definitions

The first thing we need to do is define what we mean by "remote control" and "automation." *Remote control* refers to a system that allows an action in one area to be controlled from a separate location. There may or may not be interconnecting wires. *Automation* refers to a system that can operate partially or totally without human supervision and control. In practical applications, the distinction between these two concepts is somewhat blurred. The goal of both is similar. The idea is to reduce human effort. Remote control and automation offer greater convenience than direct manual control.

Simple remote control

Virtually all remote control systems are made up of three sections, as illustrated in Fig. 1-1. The control initiator is the remote
switch or switches. It is manually activated. The switching information is then transmitted via the signal path to the controlled
device. Note that in some systems, the signal path might not exist
as actual wires. The switching information might be transmitted
as light beams, sound waves, or radio waves.

The controlled device, of course, is whatever we want to control from the remote location. In most cases, an indicator of some
sort will be necessary. In most remote control applications, the
operator is not able to see the controlled device. Indicators are
used to tell the operator the current condition of the controlled
device. For example, an LED on the control panel might light up
to indicate that the controlled device is receiving power.

Remote control systems range from the simple to the complex. A simple application is treated three different ways in Fig.
1-2. A light is to be turned on or off from a remote location.

The simplest approach is shown in Fig. 1-2A. The power
cable is simply extended, so the switch can be located wherever
we want it. This is more properly called remote switching, rather
than true remote control. The power for the controlled device is
carried right to the control point.

Remote switching can certainly be useful. It is quite commonly used in simple systems. For example, in most homes the

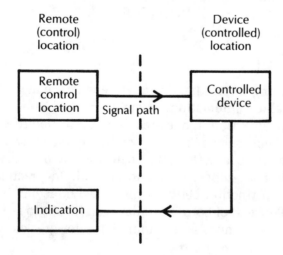

Fig. 1-1 *Most remote control systems are made up of three sections.*

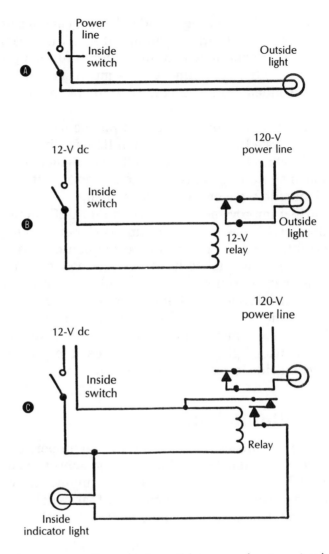

Fig. 1-2 *There are usually several possible approaches to a simple remote control application.*

outside porch light is controlled by a switch inside the house. This is a form of remote switching.

In many applications, remote switching is not practical or desirable. If the distance from the control initiator to the controlled device is very great, remote switching could be a nuisance. Heavy power cables would need to be strung between the two devices. Because of the high power flowing through the signal lines, there is always a risk of fire or electric shock.

While remote switching is perfectly adequate in some applications, in other applications you might prefer true remote control. In a remote control system, the system's full supply voltage is not carried by the signal line. Only a much smaller signal voltage is transmitted from the control initiator to the controlled device.

The most basic approach to true remote control is to use a relay, as illustrated in Fig. 1-2B. Much lighter wire is needed for the signal path. (The wire can even be eliminated altogether in some systems.) Adding multiple remote control switches is much easier than with a remote switching system.

A further improvement is shown in Fig. 1-2C. In this circuit, a latching relay is used, so a continuous signal does not need to be transmitted. A brief pulse is used to open or close the relay contacts. The relay latches itself into the new state until it receives another control pulse. This reduces power drain and lowers the signal line's requirements even further.

The circuit in Fig. 1-2C also adds the refinement of an indicator light at the remote control location. The inside light is lit when the outside light is lit. There's no need to peek out a window or go outside to see if the light is on.

Simple automation

Automation eliminates the need for a human operator, either at the device or at a remote location. Many systems can be set up to perform some or all of their functions automatically, without human supervision. A washing machine that switches itself from one cycle to the next is an example of an automation system.

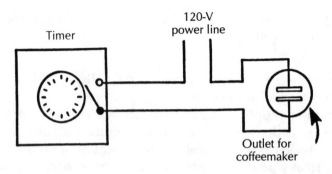

Fig. 1-3 *Many open-loop automation systems incorporate a timing device of some sort.*

Automation systems may be either open loop or closed loop. In an open-loop system, the controller pays no attention to the state of the controlled device. Many open-loop systems involve a timer. A typical example is illustrated in Fig. 1-3. Here the light will be turned on or off at the specific time preset on the timer. Notice that the timer will not check to see whether the light is already on or off. It will just send its control signal blindly at the preset time. This type of system cannot react to any failures in the system. For example, if the light bulb burns out, the timer will continue to try to turn the bulb on and off at the preset intervals.

In a closed-loop system, the condition of the controlled device is monitored by the controller. A typical closed-loop system is a thermostat, as illustrated in Fig. 1-4. The room temperature is continually monitored by the thermostat. If the temperature drops below a preset point, the thermostat tells the furnace to come on. The furnace generates heat into the room. When the temperature detected by the thermostat exceeds a specific level, the thermostat tells the furnace to shut down. In other words, the output of the controlled device (heat from the furnace) is monitored by the controller (thermostat). The controller controls the controlled device, and the controlled device controls the controller. Operation is cyclic, which is why this is called a closed-loop system.

The return monitor signal from the controlled device back to the controller is often called an error signal. The system will try to keep the error voltage at a specific fixed level and will automatically correct any deviation (error) from the standard.

In some applications, an open-loop system will do the job just fine. In others, a closed-loop system may be required. Why not just use closed-loop systems all the time? For one thing,

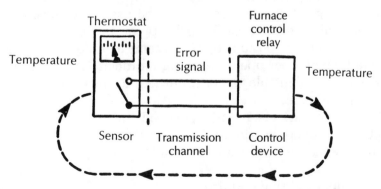

Fig. 1-4 A thermostat is an example of a closed-loop automation system.

closed-loop systems tend to be more complex and expensive than open-loop systems. If you don't need to monitor the controlled device, why bother?

Closed-loop systems must be carefully designed to prevent instability or oscillation. As an example, let's consider a thermostat in a large room. The thermostat is located at a distance from the furnace register. When the temperature in the room drops, the thermostat tells the furnace to generate more heat. It takes quite some time to heat the entire room. By the time the temperature near the thermostat is high enough for it to turn the furnace off, the area near the furnace register will be too hot. Similarly, when the furnace shuts down and the temperature starts to drop, it may be some time before the thermostat notices the change and turns on the furnace.

The whole point of a thermostat is to maintain an even temperature, but the instability of the system described here results in a room that is alternately too hot or too cold, never hitting a happy medium.

Applications for control systems

Almost anything can be a candidate for a remote control or automation system. Consider your specific needs and use your imagination. The projects in this book are designed for several specific, commonly used applications. Of course, adaptations may be made to fit the requirements of your specific applications. No law says you have to construct the project systems exactly as they are described here. Moreover, if you use your imagination and some logical thought, many of these circuits can be used in very different applications. To use one simple example, a circuit for controlling a light bulb could also be used to control a hot plate, assuming the current capability of the circuit is not exceeded. You might be able to use the same techniques I've used for an automatic door opener to design a robotic arm. You'd be surprised how adaptable the basic principles of remote control and automation are.

Of course, any electrically powered device is an obvious potential control application. Some typical electrical remote control and automation applications include

- outdoor lighting on/off
- indoor lighting on/off

- lighting level
- fans
- air conditioning
- furnaces and heaters
- stereo and radio on/off
- stereo and radio volume
- TV on/off
- TV volume
- TV, stereo, and radio station select
- intercom systems
- burglar and fire alarms
- electric coffee pots, ovens
- water heaters.

Clearly, this list could go on and on.

Less obvious applications include the physical manipulation of objects. Such applications include

- garage door opening/closing
- other door opening/closing
- window opening/closing
- draperies opening/closing
- locks
- sump pumps
- lawn sprinklers.

Let your imagination be your guide. What would you like to accomplish from a remote location or automatically? For the best results, you should design your control system to maximize flexibility. A good design permits you to add new features indefinitely. You're never boxed in.

Start out with something relatively simple. Just control one or two specific functions at first. Later you can add as many functions as you like and can afford. If you start out planning a system that is too comprehensive, you're almost sure to run into trouble. The initial expense may be inordinately high. The project may be too intimidating taken as a whole and may never be completed. The more complex a system is, the more likely it is that mistakes will creep in. On the other hand, if you start simple and work on one function at a time, you can make sure it is working perfectly before moving on to the next function. Don't complicate your life. The point of remote control and automation is to make things easier on you.

This advice is not trivial. Most experimenters (including myself, I must confess) tend to get stars in their eyes and can get themselves in deeper than expected. I have known several experimenters whose grandiose schemes have died in the throes of their own overcomplexity. It's happened to me on a few occasions. I'd like to think that I've learned my lesson by now, but, to be perfectly honest, I doubt if star-prone eyes are ever fully curable. We can all stand periodic reminding of the KISS formula—keep it simple, stupid!

Mechanical devices

Electrical switching and even changing voltages and resistances are fairly obvious types of control. Any operation that is controlled by electrical signals is probably a good choice for remote control or automation.

But other operations may not be electrically controlled. Suppose we want an automatic door opener. The door itself is not electrically controlled. Some kind of electrical to mechanical energy converter is needed.

Most readers of this book will probably face their greatest problems with the mechanical interfaces. The electronic circuitry generally isn't too complex, but the mechanical devices used in some control systems may be quite unfamiliar to the average electronics experimenter. Therefore, in this section, we will examine the basics of mechanical devices in some detail.

The first step in designing a mechanical system is the same as in designing an electronic circuit—define the purpose. Exactly what do you want the system to do? Ask yourself a series of questions. What do you want moved? How heavy is it? How far should it move? How fast should it move? How much power will it take to achieve the desired movement? Clearly, to answer these questions meaningfully, we will need to use some standard units of measurement.

"What do you want moved?" No quantitative measurement is needed for this question. It is purely descriptive.

"How heavy is it?" Any standard units of weight can be used, as long as they are consistent. We'll use the pound (lb) as our standard, although grams (g) could be used as well.

"How far should it move?" Again, we have a choice of standards for distance—the metric system or the English system. We will use the foot (ft) as our basic standard.

"How fast should it move?" Speed is a comparison of distance over time. We will define speed in terms of feet per second (ft/sec).

"How much power will it take to achieve the desired movement?" This is probably the most significant question, and the one most people are likely to be at a loss to answer.

Let's consider a simple example. We need to lift a 20-lb weight straight up. Clearly we must exert a 20-lb force in the upward direction. But this doesn't tell us how much energy or power is required for the task. We need to know how far up the weight is to be moved. Certainly more energy will be needed to lift the weight 10 ft than to lift the same weight 5 ft.

Remember that speed is defined by the distance:time ratio. Similarly, mechanical energy or force can be measured by using a weight:distance ratio. Force is measured in foot-pounds (ft-lb). The formula is simple enough:

$$\text{Force} = \text{weight} \times \text{distance} = \text{pounds} \times \text{feet}$$

To lift our 20-lb weight 10 ft, we need $20 \times 10 = 200$ ft-lb of energy.

Power includes speed, in addition to force. Obviously, it will take more energy to move that 20-lb weight 10 ft in 1 second than in 90 seconds. This time we can use the force:time ratio. That is,

$$\text{Power} = \text{force/time} = \text{foot-pounds/second}$$

Let's return to our example. To lift our 20-lb weight 10 ft in 5 seconds. We have already determined that force equals 200 ft-lb, so

$$\text{Power} = 200/5 = 40 \text{ ft-lb/sec}$$

Foot-pounds per second is a useful and clear measurement of power. However, other standards exist. Electrical power, as you know, is measured in watts (W). Foot-pounds per second can be converted to watts, or vice versa:

1 W	=	0.7376 ft-lb/sec
1 ft-lb/sec	=	1.356 W

Great, but there's still another standard measurement of power to deal with. Motors are generally rated in horsepower (hp). To continue with the conversion formulae

1 W	=	0.00134 hp
1 hp	=	746 W
1 ft-lb/sec	=	0.0018 hp
1 hp	=	550 ft-lb/sec

All mechanical devices of the type we are dealing with here are technically "machines." In this context, the word machine has a slightly different meaning than we are used to. A *machine* is a device that transforms the magnitude or direction of a mechanical force. By this definition, a computer, for example, is not a machine.

In working with any machine, it is vital to always remember the law of energy conservation. All of the energy in a system must come from somewhere, and all of the energy in a system must go somewhere. Energy can neither be created nor destroyed.

The ratio of the force exerted by a machine to the force applied to it is called the *mechanical advantage*. This is an important concept, as you will soon see. Mechanical advantage can be positive or negative (mechanical disadvantage).

One of the simplest machines is the lever. As shown in Fig. 1-5, a lever consists of three basic parts:

- power arm (input)—energy is applied to the machine here
- fulcrum—the lever is supported here
- load arm (output)—energy is exerted by the machine here.

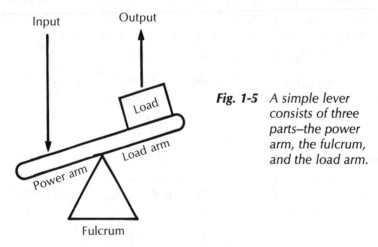

Fig. 1-5   *A simple lever consists of three parts—the power arm, the fulcrum, and the load arm.*

We can divide the lever itself into two sections. The distance from the input to the fulcrum is the power arm, or force arm. The distance from the fulcrum to the output is the load arm, or weight arm.

The relative lengths of the power arm and the load arm determine the mechanical advantage:

Mechanical advantage = power arm length/load arm length

For example, let's say the power arm is four times the length of the load arm, then

Mechanical advantage = 4/1 = 4

The force at the output will be four times the force at the input. But isn't this a violation of the law of energy conservation? No. There is a price to be paid for the increase in force. The input must be moved four times as far as we want the output to move. Since power is the ratio of force and distance, we can see that the power at the input is the same as the power at the output. (Friction and other losses are being ignored here for simplicity.) Some of the input distance has simply been transformed into output force.

Usually when we think of a lever we think of a first-order lever. This is the type shown in Fig. 1-5. The fulcrum is between the input and the output. This type of lever also performs a direction transformation. The output moves in the opposite direction as the input. The first-order lever is used to magnify the force used to move the output object, or change the distance the output object moves.

Other types of levers are also possible by rearranging the position of the various components. In a second-order lever, as illustrated in Fig. 1-6, the output is between the input and the fulcrum. The force at the output is always greater than the force at the input with this type of lever. Of course, this means, the input must always move a greater distance than the output. In a second-order lever, input and output motion are always in the same direction.

There is a third possible arrangement. You've probably already guessed that it is called a third-order lever. This type of lever is illustrated in Fig. 1-7. This time the input is between the

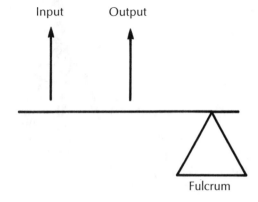

Fig. 1-6 *In a second-order lever, the output is between the input and the fulcrum.*

Output   Input

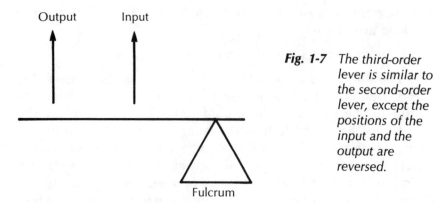

Fulcrum

Fig. 1-7 *The third-order lever is similar to the second-order lever, except the positions of the input and the output are reversed.*

output and the fulcrum. It operates in the opposite manner as the second-order lever. The input force is always greater than the output force, but the output always moves a greater distance than the input. As with the second-order lever, the input and the output of the third-order lever always move in the same direction.

Notice that physically second-order levers and third-order levers are identical. The only difference is the positions of the input and the output.

Another simple machine that is frequently useful in control applications is the pulley. A typical pulley application is illustrated in Fig. 1-8. This system is used to change the direction of force. A downward force on one end of the cord causes an upward force on the other end.

Pulleys can also be used to magnify a force. This is again thanks to the law of conservation of energy. In a pulley system the

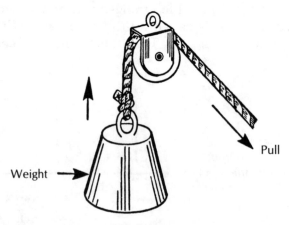

Pull

Weight

Fig. 1-8 *A pulley is another simple machine that is frequently useful in control applications.*

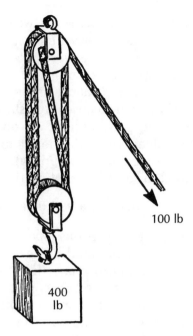

Fig. 1-9 *Multiple pulleys can be used to create a mechanical advantage.*

100 lb

400 lb

tension, or force, is the same along the entire cord. Figure 1-9 shows how multiple pulleys can be used to create a mechanical advantage. Since the tension is the same throughout the cord, the four strands of cord supporting the load will have approximately four times the force being exerted on the single-strand input. (Friction losses will waste some of the output force, but in most cases, these losses will be minimal and can be ignored.)

In practical mechanical systems, certain losses must be considered. Two of the most important factors are friction and inertia. As one object is moved against another object or substance, some of the energy will be consumed as heat at the point of contact. This is called friction. If you roll a ball down a very long, smooth hallway, it will eventually come to a stop, even if it doesn't hit anything. This is due to friction.

Friction exists in all mechanical systems outside an absolute vacuum. There is even friction against the surrounding air, which is the main reason why an arrow fired over an open field will fly just so far before gravity takes over and it falls to the ground.

In many control systems friction won't present a significant problem. In others, it will make the system require an increase in the supplied power to make up for the lost energy. In some systems, however, friction can be a very significant factor, limiting motion, causing premature wear of parts, or possibly even creating a fire hazard (waste energy is converted to heat by friction). Methods for dealing with potential problems will be discussed with the appropriate projects.

Inertia is a physical property that can be considered somewhat similar to electrical resistance. Mechanical systems tend to resist changes in motion. An object at rest tends to stay at rest, and an object in motion tends to resist any changes in speed or direction (including stopping). If an object is at rest, the force required to set it into motion (ignoring friction) is

Force = mass of object × desired acceleration

Lower accelerations (speeds) require less power. This allows the use of smaller (and less expensive) motors and other components. A slower system will tend to be less trouble-prone, with fewer snarled or broken cords, jammed motors, or other problems. In most home control applications high-speed operation is not a priority. Why complicate your life by trying to make things move faster than absolutely necessary?

Using motors

Most mechanical energy in control systems will probably come from some sort of motor. A motor, of course, is yet another type of energy transformer. It converts electrical energy (voltage) into mechanical energy as rotary motion.

Motors always produce rotary motion (the armature spins), and the speed of rotation will usually be too high for practical control applications. Speeds in a control system should usually be rather slow to increase safety and to reduce the power requirements. In this section we will explore means of transforming the high-speed rotary motion of a motor into useful mechanical energy.

One useful, but simple method for changing the speed of rotary motion is to use a pulley and belt arrangement, as shown in Fig. 1-10. Notice that the pulleys have significantly different sizes. The relative diameters indicate their relative speeds of rotation. For example, if the small pulley is attached to the shaft of a motor, the larger pulley will turn at a significantly lower rate.

The exact ratios can be easily calculated. The amount of speed reduction is inversely proportional to the pulley diameters.

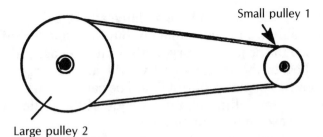

Fig. 1-10 *A pulley and belt system can be used to change the speed of rotary motion.*

That is,

$$rpm1/rpm2 = D2/D1$$

where

rpm1 = speed of pulley 1,
rpm2 = speed of pulley 2,
D1   = diameter of pulley 1, and
D2   = diameter of pulley 2.

Let's say we have a motor that has a rotary speed of 100 rpm. A 1-in. diameter pulley is connected to the motor's shaft. The other pulley has a diameter of 4 in. We now know three of the values in the equation. It just takes some simple algebra to find the fourth:

100/rpm2   = 4/1
100/rpm2   = 4
100   = 4(rpm2)
100/4   = rpm2 = 25

The small pulley will make 100 complete revolutions in a minute, while the larger pulley will make just 25. The larger pulley rotates at one-fourth the rate of the small pulley.

In designing control systems, we will usually know the motor's speed and the desired speed, and will need to find appropriate pulley sizes. For instance, a motor might have a speed of 500 rpm. We need to slow this down to 100 rpm. The first step is to arbitrarily select a size for pulley 1. Let's use 1-in. again, just because it happens to be a convenient value. Once again, we just plug the known values into the equation:

500/100   = D2/1
500/100   = D2
5   = D2

The second pulley should have a diameter of 5 in.

It probably won't ever come up, but the same technique can also be used to speed up rotary motion. In this case pulley 1 (the one attached to the motor shaft) would have a larger diameter than pulley 2. The two pulleys should be carefully aligned to prevent the belt from climbing up one side of the pulley and slipping off or wearing out prematurely.

Belts and pulleys are a good method for transferring mechanical energy from one part of a system to another. The belt is flexible enough to absorb shocks, smoothing out the force applied to the motor. Another advantage is that if something jams up the system, the belt will probably slip or break. That might not sound like an advantage, but a belt is much easier and cheaper to replace than a burnt-out motor.

Belts and pulleys are great if we need the mechanical energy in rotary form (movement in a circle). But in many (if not most) practical applications, we will need linear (straight-line) rather than rotary motion. Fortunately, it is possible to convert rotary motion into linear motion (and vice versa, if desired), although it often requires some ingenuity.

One fairly simple technique for converting rotary motion into linear motion is to use a crank, as shown in Fig. 1-11. As the shaft turns, the rod connected to the crank moves back and forth. Obviously a fairly slow rotary motion is needed for most practical applications. Slow rotary motion allows the rod to move back and forth smoothly and at a reasonable speed.

Another approach is to use a capstan, or drum, with a cord wrapped tightly around it, as shown in Fig. 1-12. As the drum turns, the cord is moved linearly. This arrangement can be used in many of the same sort of applications that might call for pulleys.

A variation of the capstan approach is illustrated in Fig. 1-13. Here one end of the cord is attached directly to the drum, or spool. As the spool is turned, the cord is wound or unwound, depending on the direction of the capstan's rotation.

Motor   Crank

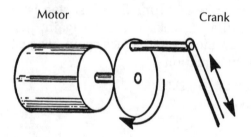

Fig. 1-11 *A simple method of converting rotary motion into linear motion is to use a crank.*

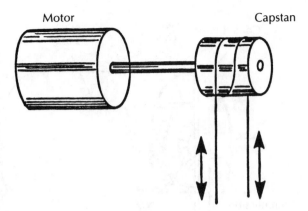

Fig. 1-12 *The rotary motion of a capstan or drum translates into linear motion of an attached cord.*

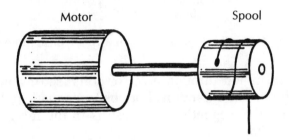

Fig. 1-13 *This is a variation on the simple capstan-based rotary-to-linear motion converter of Fig. 1-12.*

How motors work

The motor is closely related to the basic inductor and transformer. It is used to convert electrical energy into mechanical energy. This means an electrical signal through a motor can cause something to physically move. A motor is a practical application of the electromagnetic field surrounding any coil when a current passes through it.

There are many different types of motors; some are extremely tiny, while others are quite huge. Some small motors can only move very small weights, while some large motors can move tons. Some motors are designed to run on a dc voltage, and others use an ac voltage as the power source. Regardless of all of these differences, all motors are basically the same—at least in their fundamental operating principles.

A motor has two main parts: a movable coil and a fixed-position permanent magnet, or a second, fixed-position coil. An elec-

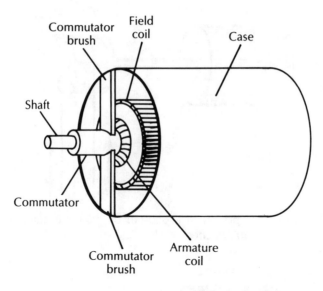

Fig. 1-14 *This is a cut-away view of a typical motor.*

trical current is fed through a set of coils, setting up a strong magnetic field. The attraction of opposite magnetic poles and the repulsion of like magnetic poles results in the mechanical motion of the motor.

Figure 1-14 shows a simplified cut-away diagram of a typical motor. Notice that there are two sets of coils in this motor. One is stationary and is known as the field coil. The other coil, which is called the armature coil or simply the armature, can freely rotate within the magnetic field of the field coil. The motor's shaft is connected directly to the movable armature coil. As the armature coil moves, the motor shaft rotates.

The commutator reverses the polarity of the current with each half-rotation of the armature and the shaft. This keeps the armature coil constantly in motion, moving its magnetic poles from the like magnetic poles of the field coil.

The operation of a motor is illustrated in Fig. 1-15. In Fig. 1-15A, the armature coil is positioned so that its motor poles are lined up with the like poles of the field coil. These like magnetic poles repel each other, forcing the armature coil to rotate, as shown in Fig. 1-15B. At some point, the attraction of the unlike magnetic poles will take over, pulling the armature into the position shown in Fig. 1-15C.

The commutator now reverses the polarity of the current, so once again the like magnetic poles of the armature coil and the

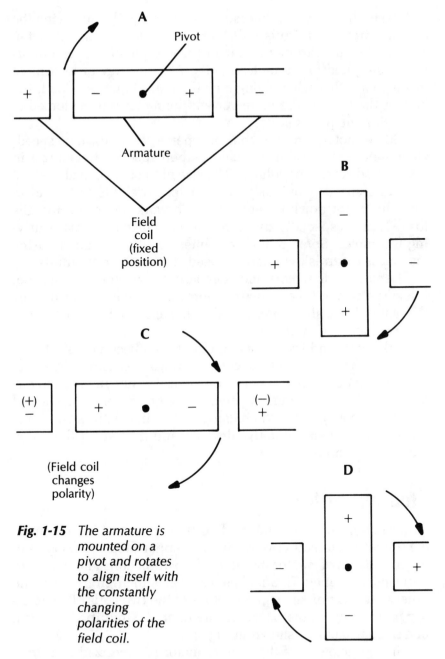

Fig. 1-15 *The armature is mounted on a pivot and rotates to align itself with the constantly changing polarities of the field coil.*

field coil are lined up, repelling each other. The whole process repeats for the second half-rotation, as illustrated in Fig. 1-15D, bringing us back to the original position of Fig. 1-15-A. The commutator reverses the current polarity again and a new cycle begins.

Assuming that all other factors remain equal, increasing the current through the coils will increase the torque of the motor. That is, a motor can move a larger (heavier) load if a larger current is supplied to the motor's coils. Another way of looking at this is to say that when using a given motor to move a load, the heavier the load is, the more current the motor will be forced to draw from the power supply.

Some motors are designed to operate at a constant speed. Other motors are designed to change their speed with changes in the applied current or voltage. The size of the load can also affect a motor's speed. Obviously, heavier load weights will tend to slow the motor down because it has to work harder to turn the load. This is especially true if a constant current source is driving the motor. Some motors are intentionally designed so the load mass controls the rotation speed, within specific limits.

Generally, dc motors and ac motors are not interchangeable. Using the wrong type of power source could damage or destroy the motor. Excessive loads can also damage small motors under some operating conditions.

Like coils and transformers, motors can become defective if there is a short in one of the coil windings. Occasionally, a broken wire in one of the motor's coils could result in an open circuit condition. Of course, either an open motor or a shorted motor will not operate. When using motors in electronic circuits, remember, they are inductive devices, and thus affect the impedance of the circuit.

Stepper motors

A special type of motor that is of particular interest in the remote control and automation fields is the stepper motor. This type of motor consists of two (or more) fixed-position coils and a pivoted permanent magnet. This magnet can rotate to align its magnetic poles with the unlike magnetic poles of the two coils. This rotating permanent magnet is called the rotor. A simplified diagram of a stepper motor is shown in Fig. 1-16.

In use, one coil of the stepper motor is energized at a time. This causes the rotor to move into any of four discrete positions, spaced 90 degrees apart. If both coils are simultaneously energized, we get four more discrete positions for the rotor, each 45 degrees away from the first set. In other words, the rotor has eight discrete positions, or steps, per rotation cycle.

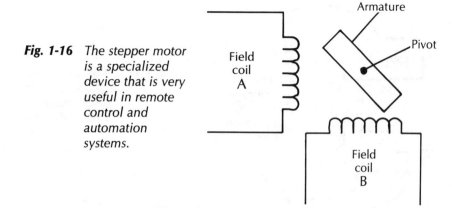

Fig. 1-16 *The stepper motor is a specialized device that is very useful in remote control and automation systems.*

Most practical stepper motors have more than two sets of fixed coils, offering a wider variety of possible rotor positions (steps). A typical practical stepper motor has a step angle of 1.8 degrees. This means that there are 200 possible step positions for the rotor.

Stepper motors are used for precise positioning of a load. A typical application is to adjust the directional position of an outdoor television or radio antenna from inside the house. The possible applications for stepper motors in robotics and other electromechanical remote control and automation applications should be fairly obvious. Almost any time we need to electrically control the precise positioning of any physical object, a stepper motor can be used.

Sensors

Most practical home control systems will require some sort of sensor mechanism to monitor the controlled device. There are countless types of sensors. Almost any condition can be electrically sensed. Of course, electrical signals can be sensed directly. In many control applications we may only need to determine whether or not the supply voltage is currently reaching the controlled device. Obviously, this can be sensed with a simple circuit that turns on (or off) when the voltage is present. For example, a relay can be used as a simple voltage sensor.

It isn't much more complicated to sense for a voltage that exceeds a specific preset limit or strays outside a specific range of acceptable values. Current flow can also be easily monitored.

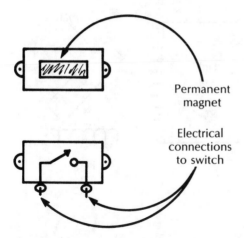

Permanent
magnet

Electrical
connections
to switch

Fig. 1-17 *Magnetic reed switches are very useful sensor devices in many control applications.*

Things get a bit more tricky when we want to electrically monitor nonelectrical conditions. Fortunately, the problem is rarely insurmountable or even unduly difficult.

Many control systems involve mechanical motion of some kind. This means that physical positions often need to be sensed. Generally, the easiest approach is to use a special mechanical switch. There are several types of switches suitable for position sensing.

A magnetic reed switch can be used to indicate proximity. They are often used in burglar alarm systems to indicate whether a door or window is open or shut. This type of switch is illustrated in Fig. 1-17. It is in two parts. One part contains a permanent magnet. This section is mounted on the moving object (such as a door). The other part contains a reed switch that responds to a magnetic field. This section is mounted on the fixed object (such as the doorjamb). Wires connect the switch to the circuitry. When the magnet is brought in close proximity to the switch, the switch closes its contacts (or opens its contacts, depending on the specific design).

Another switch that is useful for mechanical sensing applications is the snap-action switch, shown in Fig. 1-18. A small

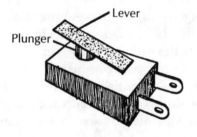

Lever

Plunger

Fig. 1-18 *Snap-action switches can be used to detect a relatively small movement.*

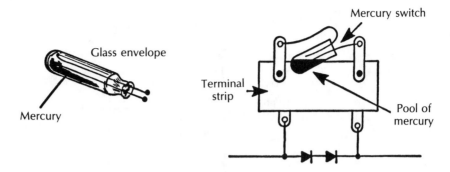

Fig. 1-19 *The mercury switch is often called a tilt switch because it can sense the angle of the switch body.*

lever touches the object to be sensed. When the object moves, it moves the lever, activating the switch. This type of switch is designed so that only a very small force is needed to actuate it.

Still another handy switch for mechanical applications is the mercury switch, illustrated in Fig. 1-19. It is basically a small glass tube with two internal electrodes that do not touch each other. A small globule of mercury is contained in the tube. If the switch is positioned so that the mercury rolls down to touch both electrodes, electrical contact is made (the switch is closed). Otherwise, the switch is open. This type of switch is useful for sensing angular position. It is often called a tilt switch.

Switches are great for simple on/off or yes/no sensing. In some control applications, we might need to monitor a continuous range of mechanical positions. Multiple switches could be used, but this is not an elegant solution and often results in undue expense and complexity.

Sometimes we can use the mechanical motion we want to monitor to turn the shaft of a potentiometer. This results in a variable resistance that corresponds to the mechanical position. A simple voltage divider circuit converts the variable resistance into a variable voltage. A simple system of this type is illustrated in Fig. 1-20.

Another continuous mechanical position sensor is shown in Fig. 1-21. The voltage drop across a forward-biased diode is more or less constant. (About 0.7 V for a silicon diode.) A string of diodes can be used as a precision voltage divider. The cord is made of two parts. The first part is an insulated cord that extends to the object being monitored (perhaps a door being opened and closed). At the end of the insulated cord is a length of uninsu-

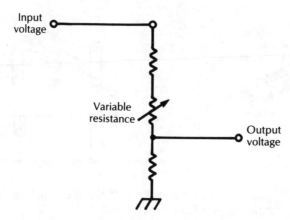

Fig. 1-20   *A simple voltage divider can convert a variable resistance into a variable voltage.*

lated conductive wire, terminating in a weight. The wire passes through a series of conductive rings at junctions between the diodes.

As the object is moved, the weight is raised and lowered. At one extreme, the wire is in contact with all the rings, shorting out

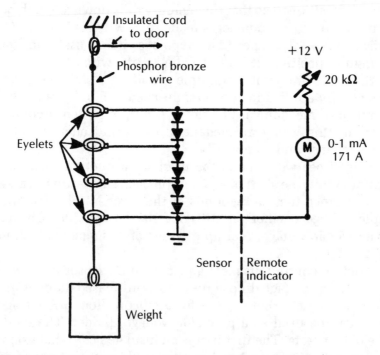

Fig. 1-21   *A string of diodes can be used to create a continuous mechanical position sensor device.*

all the diodes. As the object starts to move away from this extreme, the cord drops through the rings. The insulated section passes through some of the rings, so those diodes are no longer shorted, and their voltage drops can be sensed electrically. As the object moves further from the original extreme position, more and more of the diodes are switched into the circuit, resulting in a larger voltage drop.

Many other types of sensors are available for various special purposes. Light can be measured with photoresistors, phototransistors, or photocells. Sound can be detected with a microphone and a simple VOX (voice-operated switch) circuit. A thermistor is a component that varies its resistance in response to the ambient temperature. Several gas sensors have been put on the market to serve as electric "noses."

A crystal can often be used as a pressure sensor. This is the approach used in ceramic cartridges in inexpensive record players. Mechanical pressure along the y-axis of a crystal will cause a voltage to be generated across its x-axis. This is called the piezoelectric effect.

Another effect that can be used in monitoring applications is the Hall effect. If a current flows through a conductor under the influence of a magnetic field at right angles to the direction of current flow, a voltage drop will be produced. Hall effect magnetic sensors are commercially available.

Almost anything can be electrically sensed. In some oddball applications a little more creativity might be required because no commercially manufactured device is available for the task (at least not at a reasonable price). Various types of sensors will be used in the projects throughout this book. They will be discussed as we come to them.

Digitally controlled potentiometer ICs

A recently developed type of component that is very exciting for remote control and automation enthusiasts is the digitally controlled potentiometer IC. This device functions as an electrically controllable potentiometer (variable resistance), without any need for complex and touchy mechanical hookups. The resistance is determined by a digital input signal. Digital signals always have one of two definite and unambiguous states (high and low), so they are very convenient for electrical transmission in remote control systems of all types.

Typical digitally controlled potentiometers include the EPOT™ series by Xicor, Inc. (851 Buckeye Ct., Milpitas, CA 95035), including the X9MME, the X9MMEI, and the X9MMEM. All of these devices have the same basic structure. The pin layout of these chips is shown in Fig. 1-22. The pin functions are outlined in Table 1-1.

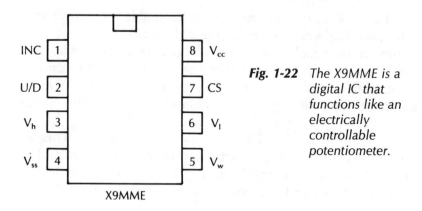

Fig. 1-22 *The X9MME is a digital IC that functions like an electrically controllable potentiometer.*

Table 1-1   Pin Functions for the Xicor X9MME Digitally Controlled Potentiometer IC.

Pin no.	Name	Function
1	INC	Increment—wiper movement control
2	U/D	Up/down control
3	V_h	High terminal of potentiometer
4	V_{ss}	Ground
5	V_w	Wiper terminal of potentiometer
6	V_l	Low terminal of potentiometer
7	CS	Chip select for wiper movement/storage
8	V_{cc}	Positive supply voltage

Basically, the X9MME contains a resistor array made up of 99 equal resistance elements. Between each resistance element and at either end of the complete string are tap points, which are accessible to the wiper element. The V_h pin (3) is connected to the upper end of the resistor array. The V_l pin (6) is connected to the lower end of this string of resistance elements. The V_w pin (5) serves as the potentiometer's wiper. It can be electrically connected to any of the internal tap points between the resistance elements. The effective position of the wiper is controlled by three single-bit digital inputs—CS (pin 7), U/D (pin 2), and INC

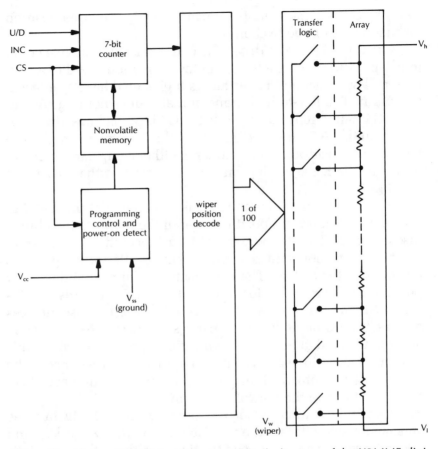

Fig. 1-23   *This is a simplified functional block diagram of the X9MME digitally controlled potentiometer.*

(pin 1). A functional block diagram of the X9MME digitally controlled potentiometer is shown in Fig. 1-23.

Pin 7 (CS) enables or disables the other two wiper control pins. A low input signal is required on this pin to enable or select the chip. On the low-to-high transition at this input, the current wiper position is stored in the X9MME's internal nonvolatile memory. A continuous high signal fed to this pin effectively disables the IC. Under this condition, nothing happens. Any data fed in at the INC or U/D inputs is simply ignored.

Pin 2 (U/D) selects the direction of the wiper's movement. This pin is active only when the CS input is low. A low signal at the U/D input causes the counter to decrement, moving the wiper down one position for each pulse on the INC input. Not surprisingly, a high signal at pin 2 has just the opposite effect. In this case, the counter increments, moving the wiper up one position for each pulse on the INC input.

Pin 1 (INC) actually drives the counter, advancing one count (one tap position) on each high-to-low transition. Neither a continuous low signal nor a continuous high signal has any effect at this input. The low-to-high transition also does nothing. Again, the INC input is active only while the CS input is low. Any input pulses on the INC input are ignored if the CS signal is high. The INC pulses can advance the counter in either an upward or downward direction, depending on the logic state at the U/D input (pin 2).

A very convenient feature of this device is that the X9MME "remembers" its wiper position, even if the power to the chip is disconnected. The last wiper position is stored in an internal nonvolatile memory and is automatically recalled when power is reapplied to the system. The manufacturer claims that the wiper position can be reliably stored for 100 years. Even if this specification should be somewhat inflated (after all, these devices haven't been around for 100 years, so it hasn't been directly tested), it is hard to imagine any practical application where this information would need to be stored more than a year or so at the most. So, for all practical purposes, the wiper position can effectively be stored in the nonvolatile memory indefinitely.

As of this writing, Xicor was offering the X9MME in three versions, reaching maximum resistances of 10 kΩ, 50 kΩ, and 100 kΩ respectively. The specifications for these three versions are outlined in Table 1-2.

Table 1-2 Specifications for the Three Versions of the X9MME Digitally Controlled Potentiometer.

Part no.	Maximum resistance	Minimum resistance	Wiper increments
X9103	10kΩ	40 Ω	101 Ω
X9503	50kΩ	40 Ω	505 Ω
X9104	100kΩ	40 Ω	1010 Ω

Clearly, the X9MME digitally controlled potentiometer can't be as finely adjusted as a true analog potentiometer. There is no way to achieve intermediate resistances between adjacent tap steps. For example, the X9103 (10 kΩ digitally controlled potentiometer) can be set for a resistance of 2,222 Ω or 2,323 Ω, but it can't be set for exactly 2,300 Ω because the resistance can be changed only in 101-Ω steps. Fortunately, in most remote control and automation applications, this shouldn't be too much of a problem. The convenience and power of being able to digitally control the resistance more than compensates for the limitations of the discrete stepped values available.

The X9MME digitally controlled potentiometer can be used in virtually any circuit using a standard manual potentiometer. Figure 1-24 shows the proper pin connections for replacing a potentiometer with an X9MME digitally controlled potentiometer IC. The power supply and control pins are not shown here.

Fig. 1-24 *The X9MME digitally controlled potentiometer can easily be substituted for any standard manual potentiometer.*

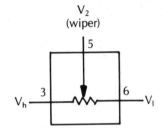

This device can be used over a fairly wide analog voltage range of at least ± 5 V. The resistance tolerance from end to end (from V_h to V_l is $\pm 20\%$. Any deviation from the nominal resistance value will be equally reflected in the individual resistance steps. For example, if the 10-kΩ X9103 is 7% low, its maximum resistance will be 9,300 Ω, instead of 10,000 Ω, and the step value will be 93.93 Ω, instead of 101 Ω. This chip is designed to operate from a standard $+5$-V power supply.

The differences in Xicor's various digitally controlled potentiometer series (X9MME, X9MMEI, and X9MMEM) lie primarily in their rated operating temperature range. For most hobbyist applications, especially indoors, these differences will be irrelevant, although some outdoor applications may require the wider thermal range of the X9MMEI or the X9MMEM. Incidentally, the "I" suffix stands for "industrial grade," and the "M" suffix stands for "military grade."

At the time of this writing, the X9MME series of digitally controlled potentiometers are included in Xicor's current catalogs and data books. I have also seen ads from other manufacturers offering similar devices. However, electronics is a rapidly changing field, so there is always a possibility that any specific component I discuss may be discontinued by the time you read about it. It is always a good idea to make sure that you can find a source for all critical components before investing money in any electronics project.

Chapter summary

In this chapter I have tried to give you a very quick introduction to basic remote control and automation techniques. This discussion has been rather general and simplified. However, most of the important concepts we will be using in the projects have been presented.

❖ 2
Building and customizing the projects

REMOTE CONTROL AND AUTOMATION PROJECTS ARE FREQUENTLY somewhat different in their construction requirements than most electronics projects. Therefore, this chapter will give you a few tips on putting together practical home control projects. In addition, every home control application is different from every other home control application. Some degree of customization is almost inevitable. This chapter will also give you some suggestions on customizing the projects from this book to suit your own individual applications.

Safety precautions

Many of the projects described in this book involve ac house current. I cannot possibly overstress the importance of safety precautions in these projects. Taking shortcuts could be fatal! There is absolutely no excuse for that.

Use proper housings for all ac-powered circuits. Never leave hot circuitry exposed during operation. Insulated housings are a must. Never use a metal housing as a termination for either end of the ac line. Use a proper ground. All wires carrying ac voltages must be properly and thoroughly insulated. Wrap any exposed junctions in several layers of black electrician's tape. Make sure it is really electrician's tape, and not just black plastic tape. Always remember, it is better to have too much insulation than to risk having too little.

Use wires that are heavy enough to safely carry the current called for in the circuit. When in doubt, use a heavier gauge wire. All ac power lines should be at least 18 or 16 gauge. Do not use 20 gauge or lighter wire for ac applications. Use strain relief devices for all external wires and cables to lessen the possibility of breakage or shorting.

Fuses are not optional! Use all fuses called for in the projects. They are there for a reason. A fuse protects both the circuitry and the operator. Leaving out a 50-cent fuse is false economy. It could be fatal. Who wants to die to save half a buck?

Never, ever substitute a higher rated fuse. This defeats the purpose of the fuse. If you can't find the specified fuse value, substitute a lesser rated fuse. If it is insufficient, all you will lose is the fuse—it might blow too soon. Never use a higher rated fuse. It might not blow in time and an expensive component in the circuit might blow to protect the fuse. Also, you could run a risk of severe injury or even death. It's not worth it!

Make absolutely certain that all mechanical parts have sufficient clearance. Try to make it impossible for any passing idiot to stick his fingers where they could get crushed by the moving parts. If it is possible to do something stupid, believe me, sooner or later, someone will do it. Nothing is ever completely foolproof—fools can be surprisingly ingenious. But try to come as close to the ideal of foolproof as you can.

Sometimes a moving part must move in the same area as humans. For example, consider an automatic door. It is an extremely good idea for all moving parts to move slowly, giving people a chance to get out of the way. Where serious injury might be a real possibility, visual or audible alarms should be used whenever the device is in motion.

Always make certain that all moving parts have sufficient clearance and can't get hung up on anything. Remember, only a fool compromises where safety is concerned. Adequate safety should be your number one concern at all times.

Finding parts

Remote control and automation projects often call for some unusual parts. Specialized sensors can often be rather difficult to locate. The normal outlets for electronic components usually won't carry the necessary mechanical parts. In many cases the specific mechanical component you need may not be available at all. You might need to customize it from something designed for an entirely different purpose. As always, ingenuity is required.

I will assume that you already know how to find standard electronic components. This section will give you a few hints on how to start looking for the "oddball" parts. Surplus stores are usually a good place to start. Most cities have at least one government/military surplus outlet. The government unloads millions of dollars worth of perfectly good equipment as surplus every year. In some cases it has been used, but parts can be easily cannibalized and reused by the experimenter. Other times, the equipment is brand new and unused, but obsolete for some reason. A lot of equipment from World War II, Korea, and Vietnam is still available in the surplus market. These are often excellent buys for the experimenter. Finally, many items have no business being sold for surplus at all. I recently bought some high-quality screwdrivers that had apparently never been used for 25 cents apiece. Apparently, the government buys a certain number of screwdrivers (or whatever) each year. Whatever hasn't been used by the end of the year is sold as surplus, and then they go out and buy a new batch of identical screwdrivers for next year. It's ridiculous. But the experimenter can take advantage of this form of governmental waste.

Industrial surplus is also available from some outlets, especially mail-order surplus houses. While there are some bargains to be found in industrial surplus (particularly when a manufacturer goes out of business), government surplus usually leads in underpriced high quality.

There are advantages and disadvantages to buying surplus by mail. Mail-order houses are usually able to offer slightly lower prices because of lower overhead costs (although this is sometimes offset by shipping and handling costs). Mail-order houses are often able to stock a larger variety of items than a walk-in store. Shopping mail order does have its disadvantages. There is the inevitable shipping delay, of course. Also some mail-order houses have minimum order limits.

A lot of experimenters are overly cautious about shopping by mail for fear of rip-offs. Actually, rip-offs are relatively rare, but they do occasionally occur. Fortunately, current FTC and postal regulations on mail-order sales are quite strict. If you do run into a problem with a mail-order dealer, contact the local postmaster.

The biggest disadvantage of buying surplus items by mail order is that you can't actually see them before you buy them.

Catalog and flyer description can be misleading. This isn't always an indication of criminal intent. It is often very hard to adequately describe an item. In some cases even photographs may not tell you enough. This is particularly a problem when you intend to cannibalize a piece of equipment for needed parts, or if you intend to use it for something other than its original purpose. Some surplus items are "whatzits." You can't use them for their original purpose because there's no way to determine what the original purpose was. Once I bought some round circuit boards that were apparently part of a missile guidance system. Who cares? I got several expensive components off of them. Buying "whatzits" by mail order is a true "pig in the poke;" it may be a great bargain or it may be worthless junk.

Buying surplus by mail is admittedly a gamble. Fortunately, if you don't buy things completely blind you should get more bargains than duds. Let the buyer beware. Addresses for mail-order surplus houses can be found in the ads at the back of experimenter-oriented magazines.

Old household appliances can be another good source of parts, especially mechanical assemblies. Think about each of the functions an appliance performs. Often these functions can be turned to other ends. Try not to consider the appliance's primary functions so much as its secondary functions. For example, a refrigerator's primary function is keeping foods cool. Many modern refrigerators include an automatic defrost feature which will usually have some sort of sensor or timer that can be used in many other control applications. All refrigerators have a door switch to turn the little light on and off. Such switches can be extremely useful in control projects. A thermostat is also a part of every refrigerator. A lot of useful goodies can be cannibalized from the secondary functions.

Hit the yard sales for used appliances. You can often pick up some great bargains. In many cases, the unit will need some repair to perform its primary function, but one or more of the secondary functions may be working perfectly.

Automotive junkyards and surplus dealers are another excellent source for mechanical parts. Ignore the original application, just think about whether or not this object can be used to perform the operation you have in mind. A little bit of creative shopping and salvaging can save you literally hundreds of dollars in a complete control system.

Although I do my best to ensure that the components called for in my projects will be available to my readers, I must offer some words of warning. Always remember that electronics is a rapidly changing field, so there is always a possibility that any specific component I discuss might be discontinued without warning by the time you read about it. This is particularly common for specialized ICs. The manufacturer might decide there just isn't a sufficient demand to keep the component in their line.

Unfortunately, when this sort of thing happens there is nothing that I can do about it. In writing my project books, I make my best guess about which components are likely to remain available for some time, but I'm certainly not infallible. I always share the disappointment of my readers who write to tell me they were unable to complete one of my projects because a key component is no longer being manufactured. I wish I could help these readers, but usually I can't. If the manufacturer has decided to discontinue a specific component and no one makes anything similar, then we're all stuck.

However, if you run into such a problem, especially if the device in question has just recently been discontinued, you might still be able to find it. Some parts suppliers might still have some left in stock. Often, discontinued components can be found (usually at excellent prices) in the catalogs of the surplus houses that advertise at the back of the electronics hobbyist magazines. In other words, if at first you don't succeed, keep asking around. The more suppliers you contact and the more catalogs you have handy, the better your chances of finding what you need.

To protect yourself from disappointment and unnecessary expense, it is always a good idea to make sure that you can find a source for all critical components before investing any money in any electronics project. For the projects in this book, I have tried to avoid any exotic or very specialized devices that are likely to be discontinued soon, but please be aware that this is always a judgement call. For each project, I will make suggestions for possible substitutions where appropriate.

I really don't think you'll have any major problems finding the required parts for any of these projects, although in some cases, a little detective work and creativity may be necessary. Actually, that can sometimes be half the fun of being an electronics hobbyist.

Substituting electronic components

Occasionally it may be necessary to substitute a component value when the specified value is not available. In most cases, high-precision component values are not necessary, which means there is some leeway in substituting values.

A close standard value can often be directly substituted. For example, a 0.2-μF capacitor may be called for. If a 0.22-μF or 0.25-μF capacitor is used, there will probably be no noticeable difference in circuit operation. However, if a specific capacitor type (mylar, ceramic, electrolytic, tantalum, etc.) is called for, try to stick with the specified type. It is called for because of certain characteristics that are important in the circuit's operation. If no capacitor type is specified, you can probably use whatever is available. Ceramic discs are reliable, widely available, and relatively inexpensive, so they are generally the best choice for non-polarized capacitances.

Resistance values can also be altered somewhat with minimal or no noticeable difference in circuit operation. If an exact value is required, it will be noted in the text. Except when a precision resistor is required, only standard resistance values are used in this book. These are all readily available. Sometimes, you can use the next standard resistance value. For example, if 12 kΩ is called for, you may be able to get away with a good 10-kΩ or 15-kΩ resistor. When such substitutions are made, it is a good idea to breadboard the modified circuit first to ensure there will be no nasty surprises when you wire the permanent version. I'd say such a change won't matter about 85% of the time. Other times, it could affect circuit operation. Just test it first before soldering.

In many cases you can create a hard-to-find value by combining components that you have on hand. Resistors can be strung in series, as shown in Fig. 2-1, to create a larger resistance. The total resistance can be calculated simply by adding together the individual series values:

$$R_t = R1 + R2 + R3 + \ldots + R_n$$

As an example, let's say we have three 180-Ω resistors in series. The total resistance will be equal to

$$R_t = 180 + 180 + 180 = 540 \ \Omega$$

The total resistance for a series combination is always larger than any of the individual series resistances.

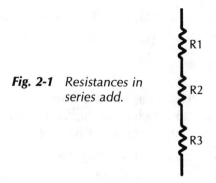

Fig. 2-1   *Resistances in series add.*

Resistors can also be combined in parallel, as illustrated in Fig. 2-2. The formula is slightly more complex for parallel resistances. The reciprocal of the total resistance is equal to the sum of the reciprocals of the individual parallel resistances:

$$1/R_t = 1/R1 + 1/R2 + 1/R3 + \ldots + R_n$$

For example, if three 100-Ω resistors are combined in parallel, the total resistance will work out to

$$
\begin{aligned}
1/R_t &= 1/100 + 1/100 + 1/100 \\
&= 3/100 \\
&= 1/33.3 \\
R_t &= 33.3
\end{aligned}
$$

The total resistance of a parallel combination is always less than any of the individual parallel resistances.

If only two resistances are combined in parallel, a slightly different formula can be used:

$$R_t = (R1 \times R2)/(R1 + R2)$$

Fig. 2-2   *The reciprocal of the equivalent value of resistances in parallel is equal to the sum of the reciprocals of the individual resistances.*

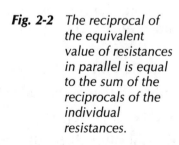
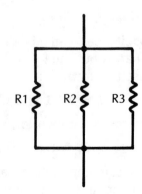

For example, if R1 = 100 Ω, and R2 = 220 Ω, then

$$R_t = (100 \times 220)/(100 + 220)$$
$$= 22,000/320$$
$$= 68.75 \ \Omega$$

If both parallel resistances are equal, the total combined value will be equal to exactly one-half the value of either of the parallel resistances. For example, if R1 = R2 = 100 Ω, then

$$R_t = (100 \times 100)/(100 + 100)$$
$$= 10,000/200$$
$$= 100/2$$
$$= 50 \ \Omega$$

Of course, series and parallel combinations can be used together, as shown in Fig. 2-3. First you'd solve for the series value of R_a and R_b (R_{ab}). Then you'd find the parallel value of R_c and R_{ab} (R_{abc}). Finally, you'd find the series combination of R_d and R_{abc} (R_t).

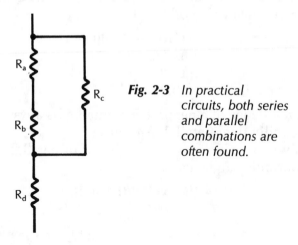

Fig. 2-3 In practical circuits, both series and parallel combinations are often found.

Capacitances can also be combined in series or parallel arrangements, but the formulas are reversed. For capacitances in series, use this formula:

$$1/C_t = 1/C1 + 1/C2 + 1/C3 + \ldots + 1/C_n$$

For capacitances in parallel, simply add the individual values:

$$C_t = C1 + C2 + C3 + \ldots + C_n$$

Often semiconductors can be substituted. Similar ICs (op amps, for example) can often be substituted by comparing their spec sheets. In some cases, identical IC devices are available in different forms. A 747 dual op amp, for instance, is two 741 op amps in a single housing.

Any time you make any kind of IC substitution, check and double check the pin numbering. Similar devices may bring different functions out to different pins. If you wire an IC wrong, there is a very good chance that it will be destroyed as soon as power is applied.

If you are not sure about whether a substitution can be made, try the circuit first in breadboard form before soldering. If you are not sure and an expensive IC is involved, I'd advise you not to attempt the substitution. You could destroy the IC.

Transistors can usually be substituted. Just use any good semiconductor substitution guide. In virtually all of the projects in this book, general-purpose transistors can be substituted. All in all, you shouldn't have much problem finding any of the electronic components for these projects.

Customizing the projects

You should certainly feel free to customize any or all of the projects to suit your own individual applications. In the section on finding mechanical parts, I said, ''Ignore the original intended application and try to determine if this object can be used for the application you have in mind.'' The same advice applies to circuitry. A circuit that responds to changes in lighting level could be made to respond to changes in temperature by substituting a thermistor for the original photoresistor. Often considerable customization can be achieved simply by changing a sensor or the input signal.

Block diagrams can be a big help. Determine what each stage in the circuit needs to do, then find a circuit that serves that function. This is far easier than designing a complex circuit from scratch. Most (if not all) complex circuits are simply made up of several relatively simple stages. As always, use your imagination.

I strongly recommend breadboarding any circuit changes before permanently soldering them. Occasionally what works on paper may not work the same way in actual practice. It's better to find out early, when it is easy to make additional changes and reuse the components, than to wait until after everything has been soldered together. Don't invite frustration.

❖ 3
Lighting

CONTROL OF ELECTRIC LIGHTING IS ONE OF THE MOST BASIC TYPES of remote control and automation applications. In many cases it would be very convenient to be able to turn lights on and off from a remote location. For example, an outdoor porch light is often controlled from inside. Automatic control of lighting is also frequently a handy thing to have. Perhaps the porch light could come on automatically at dusk, and maybe turn off at midnight.

Almost anyone could use some type of lighting control system, so we will start with this type of project. Another reason to start with lighting projects is their relative simplicity. An electrical signal needs to be turned on and off, or, in some cases, attenuated (a light dimmer). Nothing fancy is required, and there are almost never any mechanical parts to complicate the projects.

The lighting control projects in this chapter range from the very simplest to some fairly sophisticated applications. Both remote control and automation applications are covered.

PROJECT 1:
Relay switching

The most obvious type of lighting control application is to permit the lights to be turned on and off from a remote location. A very direct approach is shown in Fig. 3-1. This is more properly called remote switching rather than remote control. The power-supply voltage passes through both the controlled device (light)

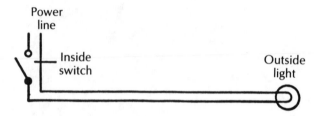

Fig. 3-1 *Remote switching is the simplest approach to "remote control," but it is extremely limited.*

and the remote switch, and all the connecting wire between them.

In some cases, remote switching is perfectly adequate. It has the advantages of being very simple and inexpensive. Virtually every home has some remote switching. If the switch for your porch light is inside, this is a form of remote switching.

However, remote switching is not always ideal. The connecting wires carry a live ac voltage. This means there is some degree of shock or fire hazard, especially if the wire is not enclosed in a wall or other structure. Fairly heavy wire with good, thick insulation must be used.

There are a great many occasions when the connecting wire will have to be strung where people might come in contact with it, or it might be strung on a path that would be far easier with a lighter and more flexible wire. It would be highly desirable if the connecting wires between the controller and the controlled device did not have to carry the full supply voltage.

The solution to these problems is a true remote control system. In a true remote control system, the connecting wire only carries a small control signal voltage. The control signal voltage is usually dc, rather than ac, which further reduces the shock hazard.

Probably the simplest approach to remote control in such an application is to use relays. An extremely simple circuit is illustrated in Fig. 3-2. The dc power source can be either a battery or a small ac-driven dc power supply. Only the small voltage and current needed to trigger the relay coil needs to flow through the connecting wires between the controller and the light. Instead of the full ac power needed to light a 100-W bulb the same job can be done by sending a small (typically under 0.5 W) dc control signal. Remote relay switching of this type can be used to control any ac-powered device, providing, of course, that the relay's switch contacts can handle the necessary power.

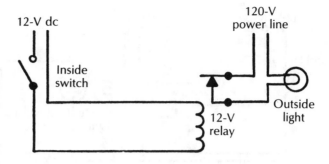

Fig. 3-2 *In a true remote control system, only a small control signal is sent from the remote controller to the controlled device.*

The simple circuit shown in Fig. 3-2 is certainly functional, but it definitely has its limitations. One problem is that there is no indicator device to tell the controller the current condition of the controlled device. Is the light already on, or is it off? Sometimes a remote indicator isn't necessary. We might be able to glance out the window to see if the porch light is on or not. However, such visibility isn't always possible or desirable.

An improved remote relay switching circuit is illustrated in Fig. 3-3. The relay in this circuit has two sets of contacts. One controls the ac light, exactly as before. The other connects the dc control signal to a third wire between the controller and the controlled location. This third wire runs back to the controller loca-

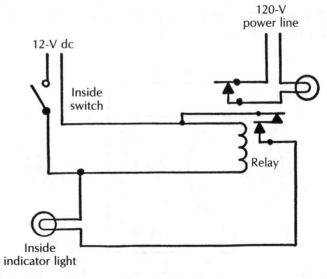

Fig. 3-3 *This improved relay circuit includes a remote indicator.*

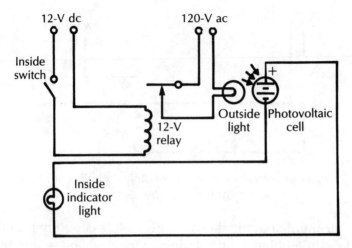

Fig. 3-4 *A better remote indication can be achieved with a true output sensor device.*

tion, to a small dc light bulb (a flashlight bulb or perhaps an LED). The remote indicator light is lit when the relay contacts are closed.

Actually, this isn't the ideal type of remote indication. All the indicator light tells you is that power is reaching the relay. It does not tell you if the outdoor light is actually lit or not. The outdoor bulb could be broken or burnt out. There could be a break in a wire past the relay.

In most simple lighting applications, this won't really be much of a problem. Just go outside and do a visual check every week or two.

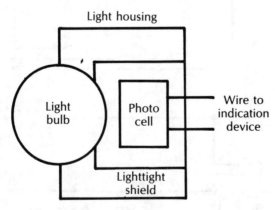

Fig. 3-5 *The photocell used as a sensor must be shielded from any external light to prevent false indications.*

If the light is critical, however, you will need a true feedback indicator to actually monitor the light itself. The circuit for one solution is shown in Fig. 3-4. A photovoltaic cell monitors the outdoor light bulb. When the light is on, the photocell generates a voltage and turns on the indicator lamp. For this system to work, the photocell must be shielded from any outside light, as shown in Fig. 3-5.

PROJECT 2:
Triac remote control

Another approach to remote switching control of lights or ac-powered appliances is shown in Fig. 3-6. A triac and a transformer are used as switching elements. The complete parts list is given in Table 3-1.

The remote switch is used to short out the secondary (6.3 V) winding of the transformer. This causes a high current to flow through the primary (120 V) winding, which actuates the triac and energizes the load. Once the triac starts conducting, current

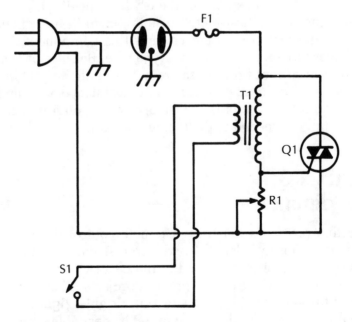

Fig. 3-6 A triac can be used for remote control of lights or low-power non-inductive ac devices.

Table 3-1   Parts list for the Triac Remote Control of Fig. 3-6.

Q1	GE-X12 (or similar) triac
T1	Power transformer—secondary 6.3 V, 1 A
R1	50-Ω, 2-W potentiometer
F1	5-A fuse and holder (3AG type)
S1	Switch

will stop flowing through the primary winding of the transformer. As a result, the winding is not likely to burn out.

Potentiometer R1 is a shunting resistor. A small magnetizing current will flow through the primary winding of the transformer, even when the remote switch in the secondary circuit is open. While quite small, this magnetizing current might be large enough to trigger the triac if the shunt resistance was not present. R1 should be adjusted for the highest resistance that will not cause false triggering.

Because of the amount of power flowing through this part of the circuit, an ordinary light-duty potentiometer cannot be used for R1. This component should be rated for at least 2 W.

To prevent damage from overheating, a heat sink should be used on the triac. A $3 \times 3 \times \frac{1}{16}$-in. cooling fin (copper or aluminum) should be adequate for the GE-X12 specified.

Note that a 5-A "fast-blow" fuse is shown in the schematic. Do not under any circumstances omit this component. Do not use a larger fuse or a "slow-blow" type fuse. Because only an extremely tiny current flows to the remote switch, very light-duty bell wire may be used between the controlled device and the remote controller. The chief advantage of this circuit is that no power supply is needed at the remote location.

PROJECT 3:
Light dimmer

While this project is not truly a remote control or automation device, it is closely related. In addition, it will be used in the remote control project described in the next section.

Lamp dimmers are always popular projects. They permit you to adjust the amount of light to suit you. Brighter light is suitable for reading, while a dimmer light might be desirable for a roman-

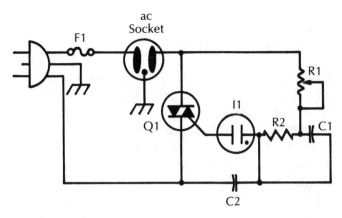

Fig. 3-7 *While not a true remote control device, this simple light dimmer circuit can be a nice addition to a control system.*

Table 3-2. Parts List for the Light Dimmer of Fig. 3-7.

Q1	RCA 40502 (or similar) triac
I1	NE-2 neon lamp
R1	50-kΩ potentiometer
R2	15-kΩ, 0.5-W resistor
C1, C2	0.068-µF, 250-V capacitor
F1	1-A fuse and holder (3AG type)

tic dinner. A simple light dimmer circuit is shown in Fig. 3-7. The parts list is given in Table 3-2.

The load (light being driven) should not exceed 400 W. A heat sink is not absolutely required for the triac in this project, but it certainly wouldn't hurt, and could prevent premature circuit failure. Potentiometer R1 is used to control the amount of power reaching the socket, and therefore, the brightness of the lamp plugged into the socket.

There is a minor compromise involved in this circuit. The neon bulb will not trip the triac until it conducts enough to turn the lamp on at a moderately bright level. Potentiometer R1 needs to be turned past this point to turn the lamp on, then it can be backed off to a dimmer glow, if desired.

This circuit should be used to adjust the brightness of electrical lamps only. Do not try to use it to control the speed of ac motors.

PROJECT 4:
Remote lamp dimmer

The lamp dimmer just described is modified for remote control in Fig. 3-8. An optoisolator is used. A small dc voltage is generated at the remote location. Potentiometer R1 controls the level of this dc voltage. The signal voltage is carried over the connecting wires to the light source in the optoisolator. The brightness of this light source is determined by the voltage fed to it.

The light source and a photoresistor are enclosed in a light-tight housing. Only light from the internal light source reaches the photoresistor. Therefore, its resistance is proportional to the brightness of the internal light source, which is dependent on the dc control voltage.

The photoresistor of the optoisolator takes the place of the potentiometer in Fig. 3-7. Otherwise, the circuit functions in exactly the same way as before, but under remote control.

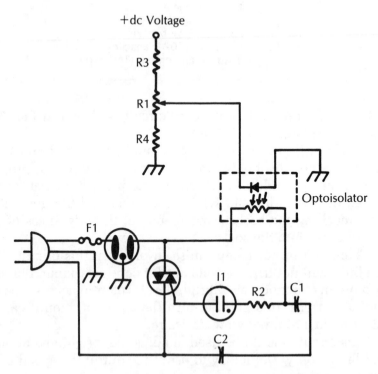

Fig. 3-8 *By adding an optoisolator, the light dimmer circuit of Fig. 3-7 can be adapted for remote control.*

PROJECT 5:
Multiple-light controller

So far the remote control projects we have described in this chapter have been designed to control just a single light or ac socket. In some applications we may want individual control over several different devices.

Of course, we could just build a number of independent circuits of the type already discussed, but that would be rather inelegant at best. It would also be unnecessarily expensive. A better solution is illustrated in Figs. 3-9, 3-10, and 3-11. The schematic is broken into three parts for clarity. The circuit is not as complicated as it might appear. The parts list for this multiple-light controller project is given in Table 3-3.

Figure 3-9 shows the remote controller section. The receiver/decoder section is shown in Fig. 3-10. Finally, the output switch-

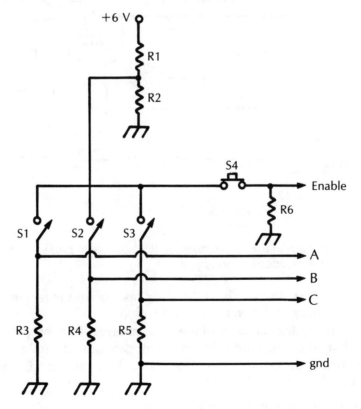

Fig. 3-9 *This is the remote controller section of the multiple-light controller project.*

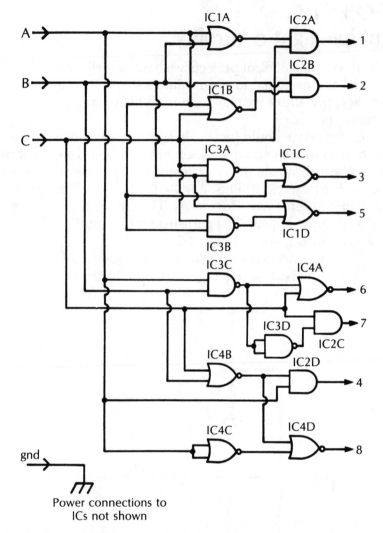

Fig. 3-10   *The digital data in the multiple-light controller project is received and decoded by this circuit.*

ing circuit is shown in Fig. 3-11. This last section is repeated for each individual ac socket to be controlled.

As the system is set up here, five control lines between the controller and the controlled devices can operate up to eight independent output devices. The signal lines are configured as follows:

- 1 = common (ground)
- 2 = enable
- 3, 4, and 5 = encoded digital data.

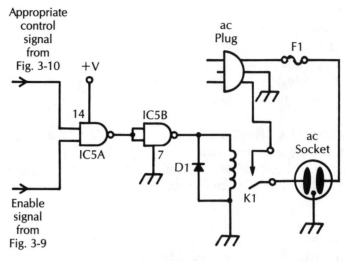

Fig. 3-11   *This circuit is repeated for each ac device to be controlled by the multiple-light controller project.*

**Table 3-3   Parts List for the
Multiple-Light Controller in Figs. 3-9, 3-10, and 3-11.**

IC1, IC4	CD4001 quad NOR gate
IC2	CF4081 quad AND gate
IC3, IC5*	CD4011 quad NAND gate
D1*	1N4001 diode
K1*	Latching relay to suit load
F1*	Fuse to suit load
S1, S2, S3	SPST switch
S4	Normally open SPST push button
R1	820-Ω resistor
R2	4.7-kΩ resistor
R3 – R6	1-kΩ resistor

*Repeat for each of up to eight controlled devices.

When an enable signal (logic 1 on the enable line) is received by the decoder, it looks at the current digital data (lines 3 – 5), and activates the appropriate latch. If the selected device was previously off, it will now be turned on. If it was already on, the new activation process will turn it off.

The eight controlled devices are controlled by three digital (binary) bits, which are encoded as follows:

- 001—device 1
- 010—device 2

- 011—device 3
- 100—device 4
- 101—device 5
- 110—device 6
- 111—device 7
- 000—device 8.

To keep the diagram simple and clear, no remote indicator devices are included in this project. It should present no real problem to add some. You can make this as simple or as detailed as your individual application demands.

PROJECT 6:
Automated guest greeter

Perhaps you don't want to leave your porch light on just in case you might have an unexpected guest after dark. But, you don't want the guest to have to wait outside in the dark until you can answer the door. The circuit shown in Fig. 3-12 offers a clever solution. The parts list for this project is given in Table 3-4. When the doorbell is rung, the porch light will turn on for a pre-determined period of time.

Ordinarily, the doorbell is activated with an SPST switch. In this project, we replace the doorbell switch with a DPST (or DPDT) switch. One section (pole) of the switch is connected directly to the doorbell in the usual manner. The second half (pole) sends a trigger signal to a timer. When triggered, the timer's output goes high, activating the output relay for a period of time determined by resistor R1 and capacitor C1. The formula for the time period is

$$T = 1.1 \, RC$$

where

T = time in seconds
R = resistance in ohms, and
C = capacitance in farads.

For this application, I think a time period of 4 to 5 minutes would probably be best. I suggest

- R1 = 1 MΩ (1,000,000 ohms)
- C1 = 250 μF.

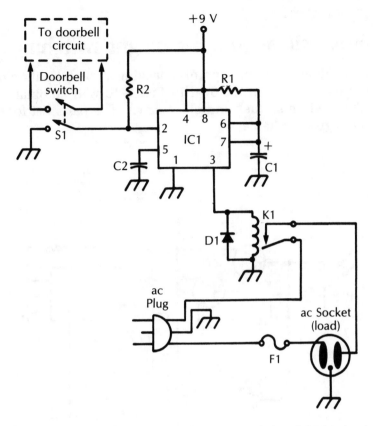

Fig. 3-12 *This circuit will automatically turn on a light when the doorbell is rung.*

Table 3-4. Parts List for the Automated Guest Greeter of Fig. 3-12.

IC1	555 timer
D1	1N4002 diode
K1	Relay to suit load
F1	Fuse to suit load
S1	Normally open DPST momentary push button (doorbell switch—see text)
C1	250-μF, 35-V electrolytic capacitor (see text)
C2	0.01-μF capacitor
R1	1-MΩ resistor (see text)
R2	10-kΩ resistor

This gives a time period of 275 seconds, or about 4.6 minutes. Of course, you can substitute other values for these components to allow different time periods.

PROJECT 7:
Photosensitive automatic light switching

Naturally, there isn't much point in leaving a porch light on during the day. The circuit shown in Fig. 3-13 will automatically turn the light on at dusk and off at dawn. The parts list for this project is given in Table 3-5.

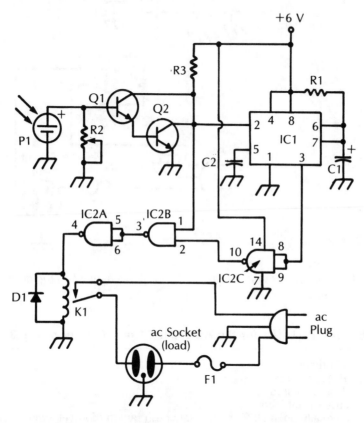

Fig. 3-13 *With this circuit lights can be turned on automatically when it is dark.*

Notice that a timer is included in this circuit. This is to prevent the light from blinking on and off in response to a passing cloud or moving shadow. The timer's delay is set for about 2 minutes (actually 112 seconds) with the component values given in the parts list (R1 = 680 kΩ, C1 = 150 μF).

**Table 3-5   Parts List for the
Photosensitive Automatic Light Switcher Circuit of Fig. 3-13.**

IC1	555 timer
IC2	CD4011 quad NAND gate
Q1, Q2	NPN transistor (2N2222 or similar)
D1	1N4002 diode
P1	Photocell
K1	ac Relay to suit load
F1	Fuse to suit load
C1	150-μF, 25-V electrolytic capacitor (see text)
C2	0.01-μF capacitor
R1	680-kΩ resistor (see text)
R2	5-kΩ potentiometer (sensitivity)
R3	100-kΩ resistor

PROJECT 8:
Sequential controller

A rather novel type of automation circuit is shown in Fig. 3-14. Up to 10 independent lights (or other ac devices) are turned on and off in sequence. Only one is on at a time.

While not suitable for applications such as the porch light situations discussed in the last few projects, this circuit can be useful, especially as an eye-catching display or warning device. Fairly short delay times should be used for such applications. The parts list for this project is given in Table 3-6.

Table 3-6   Parts List for the Sequential Controller of Fig. 3-14.

IC1	74C90 decade counter
IC2	74C41 BCD-decimal decoder
IC3, IC4, IC5	CD4049 hex inverter
C1	5-μF, 25-V electrolytic capacitor
C2	0.1-μF capacitor
R1	1-MΩ potentiometer (rate)
R2	1-kΩ resistor
R3	100-Ω resistor
D1 – D10	Diodes
K1 – K10	Relays (to suit load)

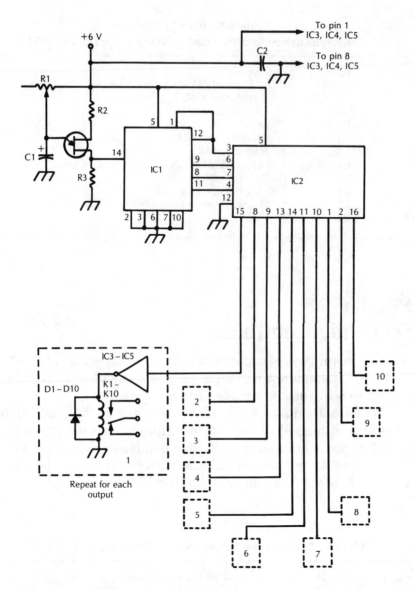

Fig. 3-14 *Eye-catching displays can be created from this sequential control-ler circuit.*

PROJECT 9:
Cross-fader

Another unusual automation circuit is shown in Fig. 3-15. The parts list is given in Table 3-7. This circuit is a cross-fader

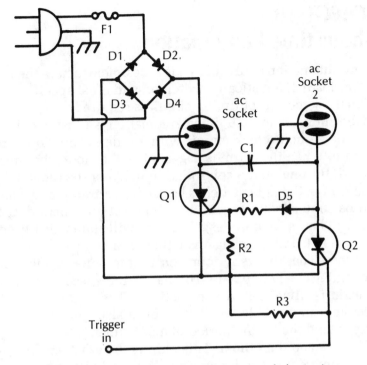

Fig. 3-15 *This is the circuit for the cross-fader project.*

Table 3-7   Parts List for the Cross-fader of Fig. 3-15.

Q1, Q2	C106B (or similar) SCR
D1 – D5	1N4003 diode
R1	47-kΩ resistor
R2, R3	4.7-kΩ resistor
C1	0.5-μF, 250-V capacitor
F1	2-A fuse and holder (3AG type)

between two lights. One light will gradually be faded down to full off, as the other one is faded up to full on.

This type of circuit should not be used for anything except lighting. Do not under any circumstances attempt to drive an ac motor with this circuit. The motor could be damaged, and it could create a dangerous situation. Also, each of the two lighting loads driven by this circuit should not exceed 100 W.

PROJECT 10:
24-hour timed automation

In many automation applications, it is desirable to have the lights (or other load) automatically switch on and off at specific times. Obviously, some sort of timing circuit is required.

One approach to such timed automation is illustrated in Fig. 3-16. Any 24-hour clock circuit with an alarm function can be used here. When the alarm is triggered by the clock, the timer is activated, turning on the controlled light. After the timer's delay period is over, the light switches back off. For example, the alarm could be set for 7:00 P.M. and the timer set for a period of 200 minutes (3 hours, 20 minutes). The light will automatically come on at 7:00 P.M. and go off again at 10:20 P.M.

A somewhat different approach to this same application is shown in Figs. 3-17 and 3-18. For clarity and space reasons, this schematic is divided into two sections. The parts list for this project appears in Table 3-8. This project also illustrates how to greatly extend the timing period of a 555 timer.

IC1 is a 556 dual timer. This chip is the exact equivalent of two 555 timers in a single housing. You could use two separate 555 chips for IC1A and IC1B, provided you correct the pin numbering. For your convenience, the pin diagram for a 555 is shown in Fig. 3-19, and the pin diagram for a 556 appears as Fig. 3-20. Also, be sure to make the power supply connections to the second 555 chip. No power supply connections are shown to IC1B in

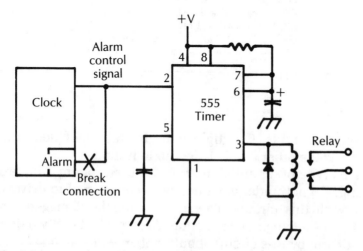

Fig. 3-16 *A 24-hour alarm clock can be used for timed automation.*

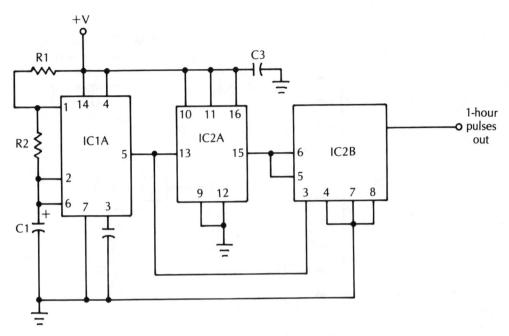

Fig. 3-17 *This circuit puts out one pulse an hour.*

this schematic because it shares its power supply pins with IC1A.

IC1A is wired as a very low frequency astable multivibrator. The formula for the timing period of this basic 555 astable multivibrator is

$$T = 0.693C1(R1 + 2R2)$$

By using the component values given in the parts list, we find that the base timing period is approximately equal to

$$
\begin{aligned}
T &= 0.693 \times 0.0005 \times (560{,}000 + (2 \times 1{,}000{,}000)) \\
&= 0.0003465 \times (560{,}000 + 2{,}000{,}000) \\
&= 0.0003465 \times 2{,}560{,}000 \\
&= 887.04 \text{ seconds} \\
&= 14.784 \text{ minutes}
\end{aligned}
$$

If the application is not too critical, you can round this off to a timing period of 15 minutes. If your particular application requires greater precision, you can use a trimpot for either R1 or R2, and more exactly set the timing period with a stop watch. Of course, when dealing with a 15-minute timing period, quite a bit of patience will be required for precise calibration. You can change the value of any of the timing components (capacitor C1

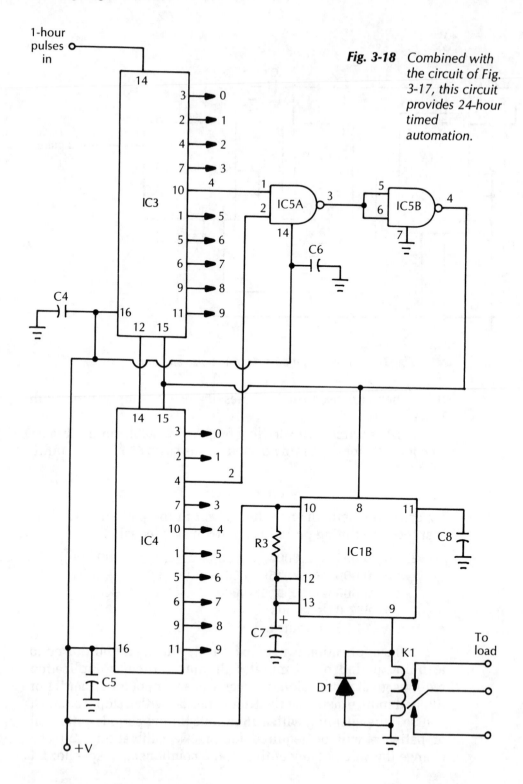

Fig. 3-18 Combined with the circuit of Fig. 3-17, this circuit provides 24-hour timed automation.

Table 3-8   Parts List for the 24-Hour
Timed Automated Control of Figs. 3-17 and 3-18.

IC1	556 dual timer (or two 555 timer ICs)
IC2	CD4027 dual JK flip-flop
IC3, IC4	CD4017 decade counter
IC5	CD4011 quad NAND gate
D1	1N4002 diode
C1	500-μF, 25-V electrolytic capacitor (see text)
C2, C3, C4, C5, C6, C8	0.01-μF capacitor
C7	100-μF, 25-V electrolytic capacitor (see text)
R1	560-kΩ, 0.25-W resistor (see text)
R2	1-MΩ, 0.25-W resistor (see text)
R3	3.3-MΩ, 0.25-W resistor (see text)
K1	Relay to suit load

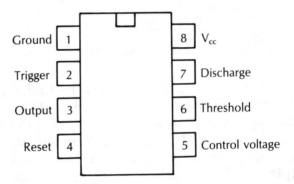

Fig. 3-19   *The 555 timer is a very useful IC for remote control and automation projects.*

and resistors R1 and R2), if it suits your individual application.

Capacitor C2, which is connected to the timer's unused voltage control input, is used simply to ensure stability. The exact value of this capacitor is not particularly critical. It might not be necessary under all conditions, but I feel it is very cheap insurance against possible frustration from circuit instability.

The 15-minute pulses from the astable multivibrator are then fed through IC2, which is a pair of flip-flops wired as simple frequency dividers. Each stage divides the input frequency by two, so together, the two stages increase the timing period to four times its original value. The output is a string of pulses separated by 1 hour. The 555 timer can be set up for direct timing periods of 1 hour, but very large component values are required, often resulting in serious stability and accuracy problems.

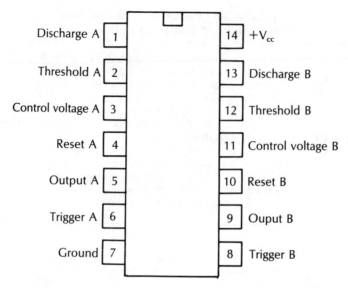

Fig. 3-20   *The 556 IC contains two 555 timers in a single housing.*

The circuitry in Fig. 3-18 extends the timing period even more. A pair of cascaded decade counters (IC3 and IC4) can divide the input pulses by any amount from 1 to 99. A simple AND gate (IC5A and IC5B) selects the desired maximum count— 24 in this case. The output of the gate is fed back to the reset inputs (pin 15) of the counters. This pulse, which occurs once every 24 hours or so (depending on the accuracy of the original timing period of IC1A), is also used to trigger a monostable multivibrator built around the second timer section of IC1B. When triggered, this timer stage goes high for a period determined by the values of resistor R3 and capacitor C7, according to this simple formula:

$$T = 1.1R3C7$$

Using the component values given in the parts list, this output pulse will last approximately

$$\begin{aligned}
T &= 1.1 \times 3{,}300{,}000 \times 0.0001 \\
&= 363 \text{ seconds} \\
&= 6.05 \text{ minutes}
\end{aligned}$$

Feel free to substitute other values for resistor R3 and capacitor C7 for different timing periods.

As long as the output of this timer stage is high, the relay (K1) will be activated. By connecting the controlled light to the normally open contacts of relay K1, the light will be turned on for a period of about 6 minutes once every 24 hours.

Capacitor C8, like C2, is just for stability insurance. Similarly, capacitors C3 through C6 protect the digital ICs from possible transients on the power supply line which could otherwise result in false triggering or other incorrect operation. Diode D1 protects the relay's coil from back-EMF. Almost any standard diode can be used here.

This circuit looks rather complicated, but as you can see from this description, it is really quite simple. The 555 and other timers will be discussed in more detail in chapter 11.

PROJECT 11:
Remote burnt-out bulb indicator

There is one problem with remote and automated lighting. Often the operator does not (or cannot conveniently) check to make sure that the lighting is actually working. The system could be working properly, but a light bulb could be burnt out, rendering the entire system useless.

How often have you seen someone driving with only one headlight? Unless someone tells them (usually a policeman, who gives a ticket along with the information), they don't know the light is burnt out.

A similar problem was one of the many contributing causes to the famous accident at the Three Mile Island nuclear power plant some years ago. The main problem was a stuck valve, but a burnt-out indicator lamp concealed the problem and confused the issue until considerable damage had already been done.

If you use a remote lighting system, you could develop a habit of periodically going out to the remote location to confirm that the light is, in fact, working. This project offers a more elegant solution. An indicator lamp (or other indicating device) is activated at the controlling station if the remote light bulb is burnt out or missing.

The schematic diagram for this project is shown in Fig. 3-21. The parts list is given in Table 3-9. As you can see, this is not a particularly difficult or complex project.

The primary of the transformer is placed in series with the remotely placed lamp. As long as the lamp is working, the circuit is completed and current flows through the transformer holding

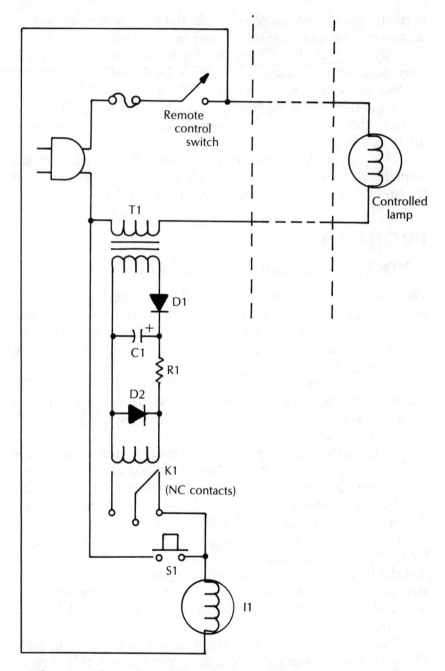

Fig. 3-21   *This is the circuit for the remote burnt-out bulb indicator.*

the relay (K1) activated. Diode D1 and capacitor C1 are a simple half-wave rectifier circuit feeding the relay's coil with a signal more or less resembling 12-V dc. Diode D2 protects the coil of the

Table 3-9   Parts List for the Remote Burnt-out Bulb Indicator of Fig. 3-21.

D1, D2	1N4002 diode (or similar)
T1	Power transformer: primary—120 V; secondary—12.6 V
K1	Relay (12-V coil) (switch contacts to suit load (small))
I1	Small indicator lamp (see text)
S1	Normally open momentary action SPST push-button switch
C1	100-μF, 25-V electrolytic capacitor
R1	1-kΩ, 0.5-W resistor

relay from burning itself out with back-EMF.

The indicator lamp (I1) is connected to the normally closed contacts of the relay. That is, when the relay is not activated, the indicator bulb will light up. When the relay is activated, the indicator lamp remains dark. If the remote lamp burns out, or if the bulb is removed, there will not be a complete circuit. No current will flow through the transformer and the relay will drop into its deactivated state, lighting the indicator lamp (I1).

Notice that the power connection for the indicator lamp is made after the remote control switch. If the indicator lamp's power was tapped off before the switch, it would glow whenever the remote light was intentionally turned off. This would serve no purpose. It would waste power and shorten the lifespan of the indicator bulb.

A normally open push-button switch (S1) can bypass the relay contacts, forcing the indicator lamp to light. This switch should be closed periodically to make sure that the indicator bulb hasn't burnt out.

If you prefer, the indicator bulb may be replaced with a buzzer or other type of indication device. The basic concepts of this project can easily be adapted to monitor many different types of remote control and automation systems.

PROJECT 12:
Automated light dimmer

This project is a variation on the basic light dimmer project introduced in project 3 and project 4. In this project, the dimmer automatically adjusts itself to a specific, manually controlled lighting level. The schematic diagram for this project is shown in Fig. 3-22, and the parts list is given in Table 3-10.

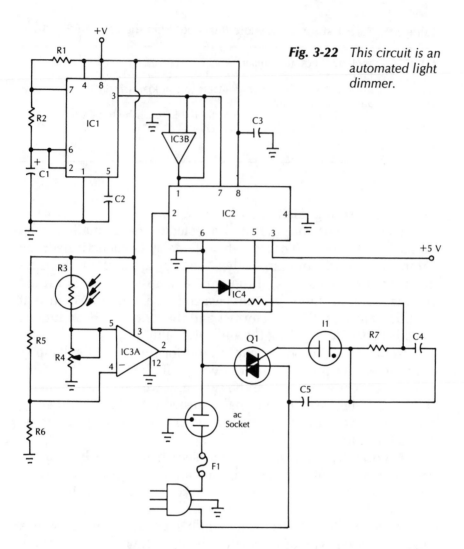

Fig. 3-22 *This circuit is an automated light dimmer.*

IC3 is one section of a quad op amp, used here as a comparator. One of the comparator's inputs is fed a fixed reference voltage derived from the R5-R6 voltage divider. Since these two resistors have identical values, the comparator's reference voltage is one-half the supply voltage. This isn't absolutely essential, but it is convenient.

The other comparator input is derived from a second, variable voltage divider, made up of photoresistor R3 and potentiometer R4. The photoresistor will change its resistance in response to the amount of light striking its surface. The potentiometer can be adjusted to control the effective sensitivity of the photoresis-

Table 3-10. Parts List for the Automated Light Dimmer of Fig. 3-22.

IC1	555 timer
IC2	X9503 50-kΩ digitally controlled potentiometer
IC3	LM324 quad op amp
IC4	Optoisolator — photoresistor output
Q1	Triac (RCA 40502 or similar), select to suit desired load
I1	NE-2 neon lamp
F1	Fuse to suit load
C1	10-μF, 25-V electrolytic capacitor
C2, C3	0.01-μF capacitor
C4, C5	0.068-μF capacitor
R1	47-kΩ, 0.25-W resistor
R2	220-kΩ, 0.25-W resistor
R3	Photoresistor
R4	1-MΩ potentiometer
R5, R6	470-kΩ, 0.25-W resistor
R7	15-kΩ, 0.25-W resistor

tor. Virtually any photoresistor will work in this type of application.

IC1 is a 555 timer wired as an astable multivibrator. Its outputs are fed to the chip select (CS) and increment (INC) inputs of IC2, an X9503 50-kΩ digitally controlled potentiometer IC. (See chapter 1 for more information on this device.)

In this particular application, the actual frequency of the astable multivibrator is almost irrelevant. I believe a moderately low frequency is best for this project. Using the component values from the parts list gives a frequency of approximately 0.29 Hz. That is, there is one output pulse every 3 seconds or so. The low-to-high transitions store the current wiper position in IC2's internal memory, while the high-to-low transitions trigger the increment function.

IC3B is another section of the quad op amp. (The remaining two sections are not used in this project.) This stage is an inverting voltage follower. It is necessary because the CS input must be low for the high-to-low transition at the INC input to be recognized.

Note, it may be necessary to add a small timing delay to either the CS or the INC signal lines. Usually the project will work fine as shown here, but apparently some X9MMEs are fussier than others about these two signals being synchronized. If you do run into such latch up problems, just add the simple

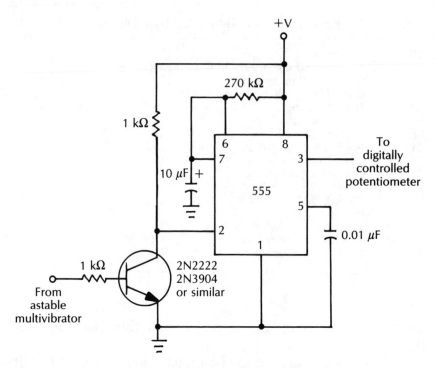

Fig. 3-23   *In some cases, this delay network might need to be added to the circuit of Fig. 3-22 to prevent synchronization problems.*

delay network shown in Fig. 3-23 between IC3B and IC2, and the project should work just fine.

    If the detected lighting level is too low, the comparator tells the X9503 to count up (via pin 2). If the photoresistor receives too much light, the digitally controlled potentiometer increments downwards. The output of the digitally controlled potentiometer determines the voltage fed to the internal LED in the optoisolator. This in turn, controls the resistance of the optoisolator's output. The remainder of this circuit is just like project 4.

PROJECT 13:
Automated staggered light balancer

This project carries the idea of project 12 one step further. Essentially, we are making two copies of the previous circuit. For convenience, the schematic is divided into two sections, shown in Figs. 3-24 and 3-25. The comparator section, illustrated in Fig. 3-24, is built just once. The control circuitry, shown in Fig. 3-25,

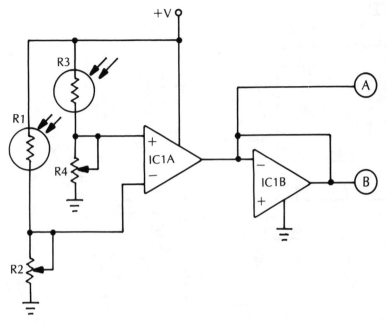

Fig. 3-24   *The comparator circuit is needed only once for the automated staggered light balancer.*

must be built twice. The part numbers and pin numbers in parentheses are for the second copy.

The two controllers drive two separate lighting units. The comparator stage matches up two photoresistors (R1 and R3). Photoresistor A measures the light from control circuit B, and

Table 3-11   Parts List for the Automated Staggered Light Balancer of Figs. 3-24 and 3-25.

IC1	LM324 quad op amp (or similar)
IC2	556 dual timer (or two 555 timer ICs)
IC3, IC5	X9503 50-kΩ digitally controlled potentiometer
IC4, IC6	Optoisolater — photoresistor output
Q1, Q2	Triac (RCA 40502 or similar), select to suit desired load
I1, I2	NE-2 neon lamp
F1, F2	Fuse to suit load
C1, C6	10-μF, 25-V electrolytic capacitor
C2, C3, C7, C8	0.01-μF capacitor
C4, C5, C9, C10	0.068-μF capacitor
R1, R3	Photoresistor
R2, R4	1-MΩ potentiometer
R5, R8	47-kΩ, 0.25-W resistor
R6, R9	220-kΩ, 0.25-W resistor
R7, R10	15-kΩ, 0.25-W resistor

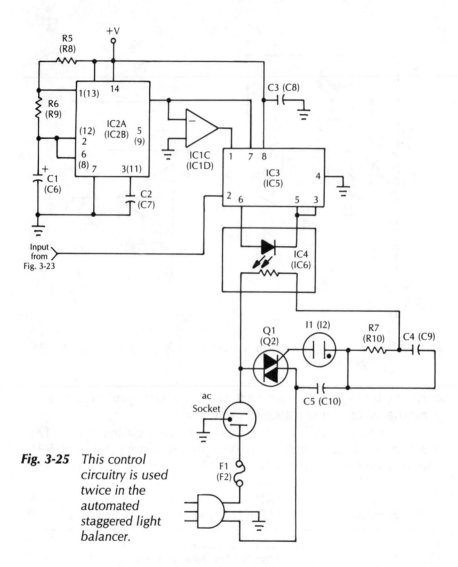

Fig. 3-25 *This control circuitry is used twice in the automated staggered light balancer.*

vice versa. The trimpots (R2 and R4) are used for sensitivity calibration and balancing the circuit.

The comparator's output is fed directly into control circuit A. For control circuit B, an inverting voltage follower (IC1B) inverts the signal, so when control circuit A is incrementing up, control circuit B is decrementing down, and vice versa. If the photoresistors are properly placed, the circuitry will automatically find a perfect balance in the levels.

The control circuit sections of this project are identical to project 12. Refer to the text of that project for more details. The complete parts list for this project is given in Table 3-11.

Doors and windows

IN THIS CHAPTER WE WILL EXPLORE THE CONTROL OF DOORS AND windows. The techniques described in this chapter can also be applied to many other applications. As always, apply ample doses of imagination.

Door and window indicators

If we are going to control an entrance (such as a door or window), we absolutely need to know the condition of the entrance. Trying to open a door that's already open or close a window that is already closed is not just futile, it can be dangerous, both to the control machinery or any nearby people. Not having an indication device for a light can be inconvenient. For a controlled door or window it can be dangerous.

There are several approaches you can take to monitor an entrance. The most basic and direct, of course, is to use one of the magnetic reed switches designed for this purpose. These switches come in two parts. One part, which is mounted on the moving door or windowpane itself, contains a small permanent magnet. No electrical connections are made to this unit. It moves with the door or window.

The other section is mounted on the doorjamb or window frame, so that it is lined up with the magnet unit when the door or window is closed. This section contains a small magnetically sensitive reed switch. Electrical connections are made to this stationary unit in the same way as for any other switch. When the

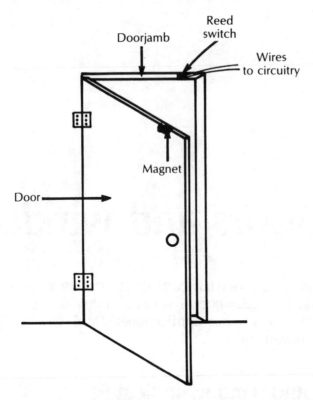

Fig. 4-1   *This is a typical installation using a magnetic reed switch to monitor a door.*

door or window is shut, the magnetic field of the moving section activates the stationary switch unit. A typical installation is illustrated in Fig. 4-1.

Magnetic reed switches are available in both normally open (NO) and normally closed (NC) versions. The "normal" condition is defined as the state of the unactivated switch away from the magnet. That is, for an NO switch, the switch is closed when the door or window is shut, or open when the door or window is opened. An NC switch works in just the reverse fashion.

These magnetic switches can be used in circuits in the same way as any SPST switch, so it is no problem to use them to activate any indication device you might choose. Magnetic reed switches are widely marketed for use in burglar alarm systems, and are available from almost all electronics and hardware dealers.

Simple switches are fine for simple "yes/no" or "open/shut" indications, but in some applications, we might need more detailed information. For example, we might need to know

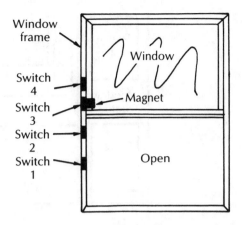

Fig. 4-2 *Multiple magnetic reed switch units can monitor multiple window positions.*

how wide a door is opened. Is it fully open? Is it half open? A quarter of the way open? Three quarters?

One solution is to mount multiple stationary switch units, so that the magnet comes into contact with each as the door or window moves through its range. A simple window installation is shown in Fig. 4-2.

Doors usually present more of a problem because most of the door's movement is away from the doorjamb. One possible arrangement is shown in Fig. 4-3. An extension arm with several switch units mounted on it extends from the doorjamb over the

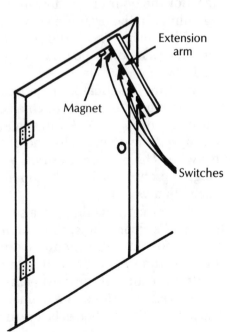

Fig. 4-3 *An extension arm over a doorway can be used to monitor various positions.*

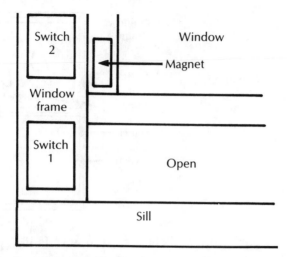

Fig. 4-4 *The multiple switch unit monitoring system runs into problems with in-between positions.*

door. This might work in some cases, but it might be highly undesirable, impractical, or dangerous in others.

If only coarse resolution is required, you may be able to get away with a multiple switch system. But if more than three or four positions are monitored, the system rapidly becomes unwieldy and expensive. In addition, in this system no allowance is made for the door or window to be positioned between monitoring points. In the example shown in Fig. 4-4, all of the switches would be open. The indicator devices would have no idea of what the window's position might be. Obviously, this is not good.

In many practical applications, a true continuous monitoring system must be devised. One convenient solution for doors is illustrated in Fig. 4-5. An insulated cord is attached to the moving edge of the door. At the end of the cord is a length of conductive wire. The phosphor bronze wire sold in hardware stores for hanging pictures will do the job nicely. The cord and wire terminate with a weight.

The wire passes through a series of conductive rings, or eyelets. As the door moves, the weight is raised and lowered. In its highest position, the conductive wire passes through all of the rings. As the weight drops due to the movement of the door, the insulated section of the cord passes through some of the rings.

Two silicon diodes are connected between each pair of rings. The entire string of diodes is forward biased. When the conductive

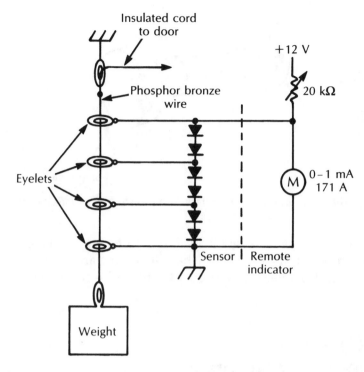

Fig. 4-5 *A diode string can be used to monitor a door's position.*

wire passes through a pair of rings, the associated diodes are out. When the insulated cord passes through one or more rings, some of the diodes become an effective part of the monitoring circuit.

The voltage drop across a forward-biased silicon diode is essentially constant—about 0.7 V. Therefore, each diode pair will drop 1.4 V unless it is shorted out. The position of the door is determined by the total voltage drop across the diode string.

Notice that intermediate positions are not such a problem with this system. The next lower position will be indicated, which should be adequate for most practical applications.

If an even more continuous indication is required, a potentiometer would probably be your best choice. The trick is to devise some arrangement for turning the shaft of the potentiometer by the motion to be monitored.

Perhaps the easiest solution for monitoring a door is to place a pulley on the potentiometer shaft, as illustrated in Fig. 4-6. The pulley is turned by a cord with one end attached to the door. The other end is terminated by a weight to keep the cord taut.

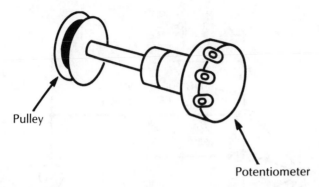

Pulley

Potentiometer

Fig. 4-6   *A pulley can be mounted on the shaft of a potentiometer.*

Other sensor arrangements for monitoring doors and windows are certainly possible. I hate to keep harping on a single point, but it is relevant throughout this book—use your imagination.

PROJECT 14:
Automatic door opener

This project can easily be adapted for many different applications just by changing the activating device. The mechanical linkage for opening and closing a door with a dc motor is illustrated in Fig. 4-7. The polarity of the voltage applied to the motor determines the direction of motion. A positive voltage opens the door, and a negative voltage closes it.

The control circuitry is shown in Fig. 4-8. When the switch is closed, the door is opened (a positive polarity voltage is applied to the motor). The door will remain open as long as the switch is held closed. Once the switch is released, the timer is triggered, holding the door open for a period determined by capacitor C1, resistor R1, and potentiometer R2. The following component values are suggested:

- C1 = 100 μF
- R1 = 18 kΩ
- R2 = 50 kΩ potentiometer.

By adjusting the potentiometer, the time delay can be anything from about 20 seconds to approximately 75 seconds.

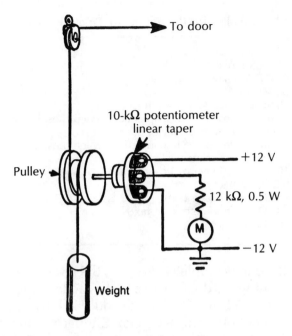

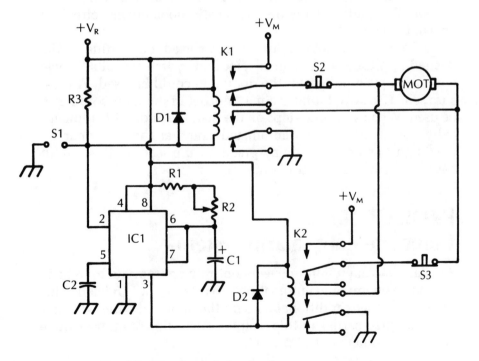

Fig. 4-7 *A pulley can be used to mechanically control a potentiometer.*

Fig. 4-8 *The control circuitry for an automatic door.*

Table 4-1   Parts List for the Automatic Door Opener of Fig. 4-8.

IC1	555 timer
D1, D2	1N4002 diode
K1, K2	DPDT dc relay to suit motor
MOT	dc Motor
S1	Normally open SPST switch (see text)
S2	Normally closed SPST snap-action switch (positioned so it opens when the door is fully opened)
S3	Normally closed SPST snap-action switch (positioned so it opens when the door is fully closed)
C1	100-μF, 35-V electrolytic capacitor (see text)
C2	0.01-μF capacitor
R1	18-kΩ resistor (see text)
R2	50-kΩ potentiometer (see text)
R3	100-kΩ resistor

After the timer completes its cycle, the reverse voltage is fed to the motor, closing the door. Notice that a diode string indicator (as shown in Fig. 4-5) is used. When the door is fully open, the positive voltage is cut off from the motor. Similarly, when the door is fully closed, no additional negative voltage is fed to the motor. This prevents damage to the motor and the mechanical linkage system. The parts list for this automatic door opener circuit is shown in Table 4-1.

Almost any switching device can be used, depending on the desired application. A remotely located switch permits remote control. A push button on the doorjamb could be used. A pressure sensitive switch under a mat in front of the door could also be used. When someone steps on the mat, the door will automatically open, just like the automated doors at the supermarket. Although I can't think of any practical application, this circuit could be activated by a timer or clock of some type.

PROJECT 15:
Complete entry alarm system

An alarm system might not seem entirely appropriate for a book on remote control and automation projects, but it is related. This project illustrates additional uses for the door and window monitors presented earlier in this chapter. Also, this circuitry can be adapted for other control projects.

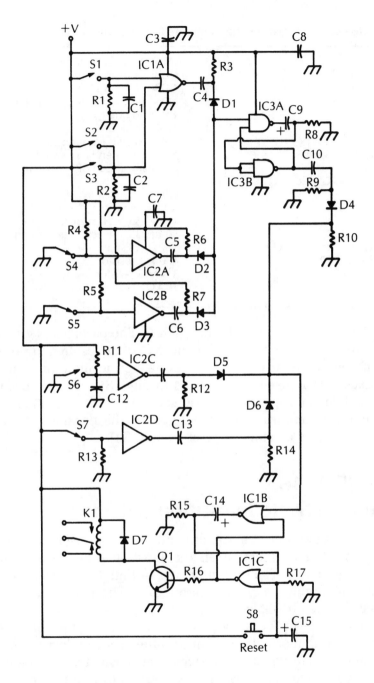

Fig. 4-9 *The sensors described in this chapter can be used for a complete entry alarm system.*

Table 4-2   Parts List for the Complete Entry Alarm System of Fig. 4-9.

IC1	CD4001 quad NOR gate
IC2	CD4049 hex inverter
IC3	CD4011 quad NAND gate
Q1	NPN transistor (2N2222 or similar)
D1-D6	1N914 diode
D7	1N4002 diode
K1	Relay to suit alarm device
S1, S2, S3, S6, S8	Normally open SPST switches
S4, S5, S7	Normally closed SPST swtiches
C1, C2, C4, C5, C6, C10, C11, C12, C13	0.01-μF capacitor
C3, C7, C8	0.1-μF capacitor
C9, C15	10-μF, 35-V electrolytic capacitor
C14	50-μF, 35-V electrolytic capacitor
R1, R2, R4, R5, R11, R13, R14	1-MΩ resistor
R3, R6, R7, R9, R10, R12	100-kΩ resistor
R8, R15, R17	3.3-MΩ resistor
R16	10-kΩ resistor

The circuit is illustrated in Fig. 4-9. The parts list is given in Table 4-2. Notice that both NO and NC switches are supported. Additional stages may be added as desired to monitor as many entrances as you need. Switches S1 through S5 turn on the alarm after a delay of approximately 30 seconds. Switches S6 and S7 trigger the alarm instantly. Once triggered, the alarm may be reset by momentarily closing the reset switch (S8).

PROJECT 16:
Sound-activated intrusion detector

Another way to protect entrances and exits is to listen for noises that shouldn't be there. Even the most stealthy intruder will make some noise. The circuit shown in Fig. 4-10 detects sound picked up through a small microphone. The parts list for this project is given in Table 4-3.

Potentiometer R1 is used to manually adjust the sensitivity of the microphone. With this control, the user can determine just how loud a sound must be to trigger the alarm. This can be used to compensate for any normal background noise in the area to be protected and to avoid false triggering.

When the detected sound from the microphone exceeds the predetermined level, a monostable multivibrator (IC1) is triggered. This timer ensures that the output alarm device is trig-

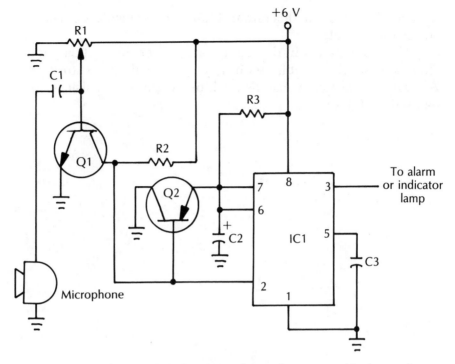

Fig. 4-10 *This circuit activates an alarm when a sound is detected.*

gered long enough to be noticed, even if the triggering sound is very brief. The timing period is determined by the values of resistor R3 and capacitor C2. Using the suggested component values from the parts list, the output goes high for about 17 seconds each time it is triggered, even if the triggering noise is just a fraction of a second long.

The output of the timer can drive an alarm bell, an electronic siren, or even a warning light at a remote location. Almost any

**Table 4-3 Parts List for the
Sound-Activated Intrusion Detector of Fig. 4-10.**

IC1	555 timer
Q1	NPN transistor (2N3904 or similar) (see text)
Q2	PNP transistor (2N3906 or similar) (see text)
C1	0.82-μF capacitor
C2	33-μF, 25-V electrolytic capacitor
C3	0.01-μF capacitor
R1	10-kΩ potentiometer
R2	1-kΩ, 0.25-W resistor
R3	470-kΩ, 0.25-W resistor

load can be driven by this circuit if a suitable relay is connected across the output.

The components in this circuit are not terribly critical. You should have no trouble with any reasonable substitutions. Almost any reasonably matched pair of low-power transistors can be used for Q1 and Q2.

❖ 5

Temperature control

REMOTE CONTROL, AND ESPECIALLY AUTOMATION TECHNIQUES, can be very useful for providing a more comfortable environment. Probably the most widespread form of environmental control is the control of temperature. A thermostat is an automatic device for controlling heating and air conditioning equipment. A thermostat can be a simple automated switch or it can be a highly sophisticated device with special features. This chapter offers several projects for automating temperature control in a variety of ways.

Temperature sensors

There are several approaches to electrically monitoring temperature. The most commonly used temperature sensors in electronic circuits are thermocouples, diodes, and thermistors. We will briefly discuss each of these in the next few pages.

Thermocouples

Perhaps the simplest electrical temperature sensor is the thermocouple. It consists of nothing more than a junction of two wires made of different metals. When the junction is heated, a voltage proportional to the applied temperature is developed across the junction. This is known as the Seebeck effect. The developed voltage can then be measured and used to control many different circuits, as desired.

Diode temperature sensors

A standard silicon diode can also be used as a temperature sensor. If a small forward bias is applied to the diode, the voltage drop across the diode will respond to changes in temperature at a rate of about 1.25 mV (0.00125 V)/°F.

A simple diode temperature sensor circuit is illustrated in Fig. 5-1. If you are familiar with electronics theory, you should recognize this circuit as a variation on the Wheatstone bridge. Wheatstone bridges are widely used in measurement circuits of many different types.

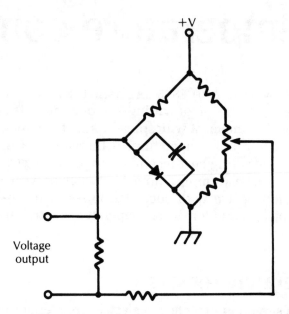

Fig. 5-1 *A diode can be used as a simple temperature sensor.*

The supply voltage for this circuit should be between +1 and +1.5 V. Almost any silicon diode can be used in this type of application, including the popular and inexpensive 1N914 diode.

Thermistor

In modern circuits, probably the most popular type of temperature sensor is the thermistor, or thermal resistor. All resistance elements are temperature sensitive to some extent. A thermistor is specifically designed to emphasize the effects of temperature on the component's resistance in a predictable manner.

There are two types of thermistors—NTC and PTC. NTC thermistors, which are the more common, have a negative temperature coefficient. This means that resistance decreases as temperature increases. PTC thermistors, on the other hand, have a positive temperature coefficient and operate in the opposite manner—resistance increases as the temperature increases. Both types can be useful in various applications.

Thermistors are very handy for temperature sensing applications, but they have their definite limitations. In most cases, temperature-resistance curves are not linear. Temperature measurement scales (Fahrenheit and Celsius), however, are linear systems. This means it is difficult to design a thermistor-based circuit that is linear over a wide range—difficult, but not entirely impossible. Many electronic thermometer circuits have been designed around thermistors. NTC thermistors can exhibit negative resistance characteristics under some circumstances, leading to thermal runaway and possible self-destruction of the thermistor.

Remember that Ohm's law specifies the interrelationships of resistance, current, and voltage:

$$R = E/I$$
$$E = RI$$
$$I = E/R$$

where

    R = resistance,
    I = current, and
    E = voltage.

Fig. 5-2 *An NTC thermistor can be susceptible to thermal runaway.*

Consider the simple series circuit of Fig. 5-2. An NTC thermistor is in series with a fixed resistor. For simplicity, we will ignore the thermal drift of the fixed resistance. We will assume the applied voltage is 10 V and the fixed resistance is 1 kΩ (1000 Ω). The resistance of the thermistor, of course, is dependent on the temperature. The higher the temperature, the lower the resistance, because this is an NTC device.

At some temperature, the thermistor will have a resistance of 10 kΩ (10,000 Ω). The total resistance of the circuit is 11 kΩ (1 kΩ + 10 kΩ). Therefore, the current flowing through the circuit will be equal to

$$I = E/R = 10/11,000 \approx 0.00091 \text{ A} = 0.91 \text{ mA}$$

The voltage drop across the fixed resistance is approximately equal to

$$E = RI = 1000 \times 0.00091 \approx 0.91 \text{ V}$$

and the thermistor will drop about

$$E = RI = 10,000 \times 0.00091 \approx 9.1 \text{ V}$$

Because of the negative temperature coefficient, the resistance of the thermistor will drop as the ambient temperature increases. At some higher temperature, the thermistor's resistance will be 1 kΩ (1000 Ω), making the total resistance in the circuit equal to 2000 Ω. Now the current flowing through the circuit is about

$$I = 10/2000 = 0.005 \text{ A} = 5 \text{ mA}$$

Note that the current increases with the temperature.

The voltage drop across the two resistances will be equal because of their equal values:

$$E = 1000 \times 0.005 = 5 \text{ V}$$

What do these two examples tell us? As the temperature increases, the current flow and the voltage drop across the thermistor also increases. So what?

Well, what happens to the voltage dropped across a resistance element? It is dissipated as heat. The higher the voltage drop, the greater the heat produced. Therefore, as the ambient temperature increases, the resistance of the thermistor drops and the voltage dropped across it increases. The voltage drop causes the thermistor to generate heat. Obviously, this increases the temperature sensed by the component, further reducing its resistance, which increases the voltage dropped across it, increasing the generated heat, and so on and so on.

As long as the current is relatively small, this self-heating effect probably won't be too terribly significant. It may result in a slight loss in accuracy, but no real harm will be done. However, when large currents flow through an NTC thermistor, the effect can be quite large, resulting in thermal runaway. The sensed temperature becomes totally meaningless, and the circuitry can be damaged from overheating.

PTC thermistors are not subject to thermal runaway. As the temperature increases, so does the resistance, thus reducing the voltage drop and decreasing the amount of heat generated. In fact, PTC thermistors are often used as temperature regulators. When power is applied, the thermistor will heat itself until a stable level is reached. A fairly constant temperature will be maintained despite any minor changes in the ambient temperature or thermal load.

Thermistors are not suited for monitoring rapid temperature fluctuations. If a thermistor is moved from one temperature extreme to another, it cannot respond instantly because of inertia. The mass of the thermistor's body must heat up or cool down to match the new ambient temperature. Obviously, this takes a finite amount of time. It should be equally obvious that the larger the thermistor's body, the greater the time required for it to respond to a change in ambient temperature. For a small disc or bead thermistor, the response time will typically be just a few seconds. Large thermistors, of course, will have a longer response time. Some heavy-duty thermistors have a body that is 1 in. in diameter. These units might require up to 2 minutes to respond to a change in the ambient temperature.

Probe assemblies

In most applications, the temperature sensing element should be housed in some kind of protective probe assembly. This is especially true if the temperature of a liquid is to be measured. Any conductive liquid (such as water) can shunt out the connecting leads, resulting in false readings. Nonconductive liquids can soften the epoxy protecting the delicate active element. Corrosive liquids can damage the leads or the sensor body.

Except for hermetically sealed glass body thermistors, liquid can creep inside the component itself, seeping through minute openings between the leads and the epoxy coating. This can result in misreadings and eventual corrosion of the sensing element.

While these problems are most prevalent when the sensors are immersed in a liquid, they can show up in other applications too. Even measuring air temperature can be problematic if the humidity is high. Also, an unshielded temperature sensor can be cooled by passing winds. This is desirable, of course, if you want to monitor the wind chill. But if you're interested in the actual temperature, it can be a significant problem. High winds or contact with other objects can damage a delicate temperature sensor.

The solution to all these problems is not particularly difficult. You simply have to house the sensor in some sort of probe assembly. A typical probe housing is illustrated in Fig. 5-3.

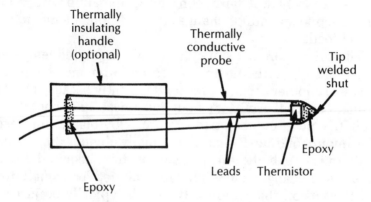

Fig. 5-3 *For most practical applications, the temperature sensor should be enclosed in a probe housing.*

The probe itself is usually a tube of glass or stainless steel, welded shut at one end. A temperature insulating handle is often placed on the opposite end of the probe. The sensor is inserted in the tip of the probe. Often a bit of epoxy is used to improve mechanical strength and improve the thermal contact between the sensor and the external probe housing.

The connecting leads are soldered to the sensor and brought out through the back of the probe. Generally, another dab of epoxy here is a good idea. This relieves strain on the connecting wires and prevents them from being pulled free of the sensor.

For most purposes, a "straight stick" probe, as shown in Fig.

5-3 will do the job just fine. For more permanent setups, a threaded probe can be screwed into a socket or holder.

PROJECT 17:
Simple electronic thermostat

A fairly simple electronic thermostat circuit for controlling an electric heating element is shown in Fig. 5-4. The parts list for this project is given in Table 5-1.

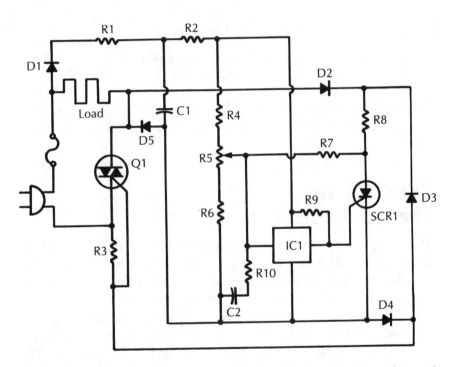

Fig. 5-4 *This simple thermostat circuit can directly control an electrical heating element.*

The heater load may be rated from 10 mA (0.01 A) up to about 35 A. The triac should be selected to handle the appropriate amount of current. Similarly, the fuse size should be determined by the heater element. With larger heater elements, household-type screw-in fuses might be used instead of the light-

Table 5-1   Parts List for the Simple Electronic Thermostat of Fig. 5-4.

IC1	LM3911
Q1	Triac (see text)
SCR1	2N5064 (or similar)
D1 – D5	1N4004 (or similar)
C1	0.5-μF, 200-V capacitor
C2	0.05-μF capacitor
R1	3.3-kΩ resistor
R2	82-kΩ resistor
R3	27-Ω resistor
R4, R6	16.5-kΩ, 2% wirewound resistor
R5	10-kΩ potentiometer
R7	10-MΩ resistor
R8	100-Ω resistor
R9	68-kΩ resistor
R10	1.5-MΩ resistor

All resistors should be 0.5 W or better.

duty 3AG types found in most electronic circuits. Select the fuse rating conservatively. It is better to use a too-small fuse than a too-large one.

PROJECT 18:
Over/under temperature alert

The circuit shown in Fig. 5-5 indicates whether a monitored temperature is within a preset range. Moreover, it also indicates whether the out-of-range temperature is too high or too low. A parts list for this project is given in Table 5-2.

Basically, this circuit is a window comparator comparing voltages from two resistive voltage divider strings. One voltage divider is made up of three fixed resistors (R1, R2, and R3). The other voltage divider is comprised of a trimpot (R5) and a thermistor (R4). The thermistor changes its resistance in proportion to its temperature. The trimpot is used to adjust the effective sensitivity of the thermistor.

IC1A compares the temperature-dependent voltage with the upper portion of the fixed reference voltage. At the same time, IC1B compares the temperature-dependent voltage with the lower portion of the fixed reference voltage. Then, IC1C compares the combined outputs of the first two comparator stages with another reference voltage derived through resistor R10.

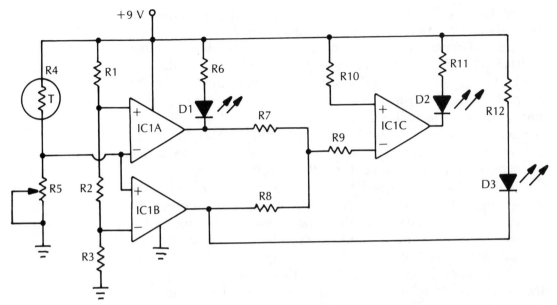

Fig. 5-5 *This circuit indicates when a monitored temperature is too high or too low.*

Table 5-2   Parts List for the Over/Under Temperature Alert of Fig. 5-5.

IC1	LM339 quad comparator (see text)
D1, D2, D3	LED
R1, R3	22-kΩ, 0.25-W resistor (experiment to change switching points) (see text)
R2	10-kΩ, 0.25-W resistor (experiment to change switching points) (see text)
R4	Thermistor
R5	1-MΩ trimpot (sensitivity)
R6, R11, R12	330-Ω, 0.25-W resistor
R7, R8	27-kΩ, 0.25-W resistor
R9	47-kΩ, 0.25-W resistor
R10	68-kΩ, 0.25-W resistor

Three LEDS (D1, D2, and D3) indicate the monitored temperature's relationship to the preset range. Only one of these LEDs should be lit at a time. As long as power is applied to the circuit, one and only one of the LEDs should light up.

The individual meanings of these LED indicators are as follows:

- D1—too high
- D2—in range
- D3—too low.

The high/low voltages fed to these LEDs can easily be tapped off and fed to other circuitry within a larger system.

The reference range is set by the values of voltage divider resistors R1, R2, and R3, along with the characteristics of the particular thermistor used. You'll probably want to experiment with alternate values for these reference resistors. In some applications, it may be desirable to replace resistors R1 through R3 with trimpots. Remember, these three resistances interact in setting the comparator switch points. Changing any one resistance changes the balance of the entire voltage divider string.

This project is designed around three sections of an LM339 quad comparator IC. The fourth section may be used in other circuitry in a larger system. If this extra comparator section is left unused, however, its inputs and outputs should all be grounded. A comparator section with floating inputs or outputs can adversely affect the operation of the other comparator stages on the same chip. Standard op amps can be used in place of the dedicated comparator stages of the LM339.

PROJECT 19:
Alternate electronic thermostat

Figure 5-6 illustrates how a comparator circuit similar to project 18 can be modified to function as a simple electronic thermostat. A parts list for this project appears in Table 5-3.

Here we are using just a single comparator stage built around an op amp (IC1A). If the monitored temperature goes above a specific preset level (determined by the values of resistors R1 and R2), the comparator's output goes high.

The output of the comparator is fed through a noninverting voltage follower (IC1B), which serves as a buffer. When the comparator's output is high, the relay is turned on. If the monitored temperature is below the reference point, the comparator's output goes low and the relay is deactivated.

We are assuming here that the thermostat is controlling some sort of heating device. If you want to control a cooling device with this circuit, simply use the relay's normally closed contacts instead of the normally open contacts, as shown in the schematic diagram. Alternately, you could reverse the positions of the thermistor and its sensitivity control trimpot.

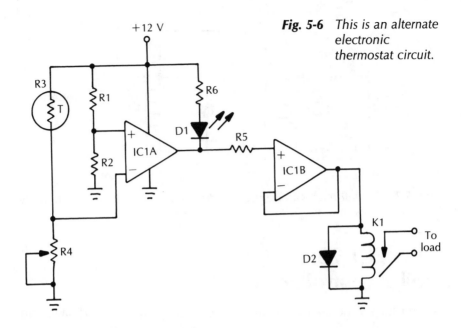

Fig. 5-6 *This is an alternate electronic thermostat circuit.*

Table 5-3 Parts List for the Alternate Electronic Thermostat of Fig. 5-6.

IC1	LM324 quad op amp (two sections only) (or similar)
D1	LED
D2	1N4002 diode (or similar)
R1	22-kΩ, 0.25-W resistor (experiment to change switching points) (see text)
R2	33-kΩ, 0.25-W resistor (experiment to change switching points) (see text)
R3	Thermistor
R4	1-MΩ trimpot (sensitivity)
R5	37-kΩ, 0.25-W resistor
R6	330-Ω, 0.25-W resistor
K1	12-V relay, contacts to suit load (see text)

Most practical heating and cooling devices draw fairly heavy currents. The buffer in this circuit might not be able to control a large enough relay directly. Figure 5-7 shows how a smaller relay can be used to control a larger relay. While the intermediate voltage source is shown here as a battery, any suitable voltage source may be used, even the supply voltage for either the control circuit or for the load, depending on the requirements of the secondary relay's coil.

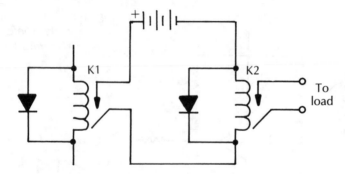

Fig. 5-7 *A small relay can be used to drive a larger relay to control hefty loads.*

PROJECT 20:
Boiler controller

If you have an oil-based water heater, this project could save you quite a bit on fuel bills. In water heaters fueled by oil, the system's water temperature is controlled by a device called an aquastat. Usually there are manual adjustments for water temperature and circulator control. Typically, the circulator is set about 20°F lower than the water temperature.

Greatest fuel efficiency can be achieved by using a somewhat lower water temperature setting in summer than in winter. Good "rule of thumb" values are about 180°F for winter and 160°F for summer. Notice that with these recommended values, the water temperature varies inversely with the outside temperature.

This project monitors the outside temperature and automatically adjusts the water temperature accordingly. For best results, the circulator control should be set to about 125°F to 135°F. Rather than settling for a seasonal approximation, the water heater will respond directly to changes in the weather, maximizing fuel efficiency.

The input section of this project is shown in Fig. 5-8, and the output section is shown in Fig. 5-9. The project is split into two sections to make the schematics a little easier to work with. The parts list for the entire project is given in Table 5-4.

Switch S1 permits the user to select between automatic and manual (override) modes. The relay's (K1) switch contacts are wired in series with the boiler's aquastat. The lower and upper temperature limits are set via potentiometer R32 and R25 respec-

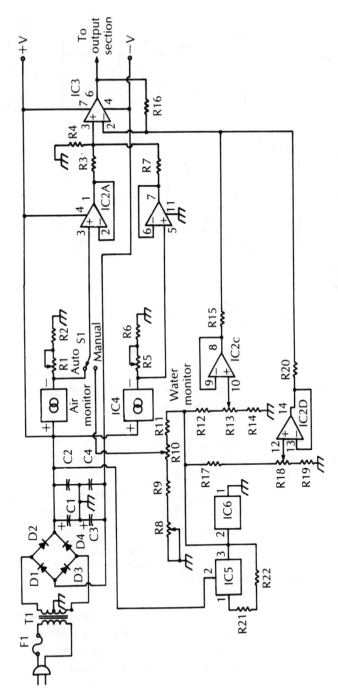

Fig. 5-8 This is the input section of an economical boiler-controller circuit.

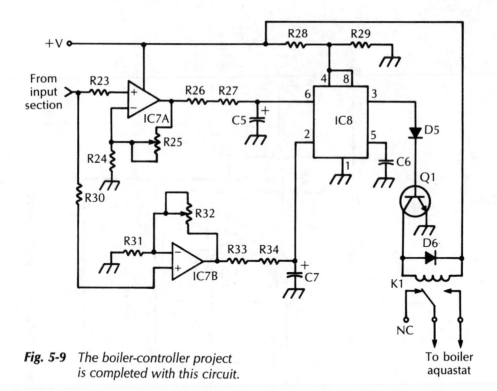

Fig. 5-9 *The boiler-controller project
is completed with this circuit.*

tively. The other trimpots are for calibration purposes. The calibration procedure is as follows:

1. Set to automatic mode.
2. Adjust R13 for 4.6 V at pin 5 of IC2.
3. Adjust R18 for 4.6 V at pin 12 of IC2.
4. Adjust R5 so that the voltage between pins 10 (–) and 5
   (+) corresponds to the measured temperature (1°F = 10
   mV) (water monitor sensor).
5. Adjust R1 so that the voltage between pins 12 (–) and 3
   (+) corresponds to the measured temperature (1°F = 10
   mV) (air monitor sensor).

These tests are best performed with both sensors at the same
temperature, along with an accurate reference thermometer.
Place the two sensors and the thermometer in a dry, even-temper-
ature location that is protected from significant passing breezes.

Table 5-4   Parts List for the Boiler Controller of Figs. 5-8 and 5-9.

IC1, IC4	AD590 temperature sensor
IC2	LM324 quad op amp
IC3	741 op amp
IC5	LM334 constant current source
IC6	6.9-V temperature-stabilized reference voltage
IC7	747 dual op amp
IC8	555 timer
Q1	NPN transistor (Radio Shack RS2017 or similar)
D1 – D4	1N4002 diode
D5	1N1202 diode
D6	1N4004 diode
T1	18-VCT 2-A transformer
F1	5-A fuse and holder (3AG)
S1	SPDT switch
K1	12-V dc DPDT relay with 160-Ω coil
C1, C3	1000-μF, 50-V electrolytic capacitor
C2, C4	0.1-μF capacitor
C5, C7	5-μF, 35-V electrolytic capacitor
C6	0.01-μF capacitor
R1, R5	5-kΩ trimpot
R2, R6, R24, R28, R31	15-kΩ resistor
R3, R4, R7, R15, R16, R20	100-kΩ resistor
R8	20-kΩ trimpot
R9	33-kΩ resistor
R10	10-kΩ potentiometer
R11, R29	12-kΩ resistor
R12, R17	4.2-kΩ resistor
R13, R18	1.5-kΩ trimpot
R14, R19	8.2-kΩ resistor
R21	27-Ω resistor
R22	4.7-Ω resistor
R23, R30	10-kΩ resistor
R25, R32	50-kΩ potentiometer
R26, R33	18-kΩ resistor
R27, R34	2.2-kΩ resistor

6. Measure the output (pin 6) of IC3. This voltage should correspond to the measured voltage × 2 (1°F = 0.2 V).

If this last test does not give the correct results, begin the calibration over. If it still does not work right, something is wrong with the circuit. Double-check all part values.

7. Set S1 for manual mode.
8. Adjust R8 until the voltage at the junction of R9 and R10 is 4.6 V.
9. Adjust R10 for an output of 1.8 V from IC3 (pin 6).

PROJECT 21:
Temperature equalizer

As Fig. 5-10 illustrates, a thermostat is a closed-loop type of auto-mation circuit. This simply means that there is a continuous, cir-cular path throughout the automation system. The thermostat monitors the room temperature and controls the furnace. The heat from the furnace changes the room temperature, which affects the thermostat. The input (room temperature) and output (heater) are interrelated, affecting each other directly. Because of this interrelationship between the input and the output, the closed-loop system can exhibit some instability or oscillation.

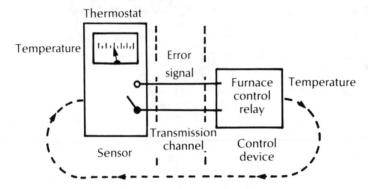

Fig. 5-10   *A thermostat is a closed-loop type of automation circuit.*

Consider what happens if the thermostat is positioned some distance from the heater duct and there is only limited air circu-lation in the room. The temperature in the room drops, so the thermostat tells the furnace to put out more heat. Because of the distance from the heater duct to the thermostat, it will be some time before the temperature at the thermostat is high enough for it to shut off the furnace. Part of the room will now be too hot.

When the furnace is finally turned off, the room will begin to cool, especially near any heat leaks, such as doors and win-dows. Let's assume these leaks are also a significant distance from the thermostat. Again, it will be a considerable time before the thermostat can sense the drop in temperature. By this time, part of the room might be too cold.

You can see how the room's overall temperature oscillates from too hot to too cold, and back, without ever reaching a happy medium. You not only don't get the proper benefits of your heating system, you will also tend to waste quite a bit of fuel or power with the unnecessary oscillations.

Of course, a partial solution is to position the thermostat close to the heating register and significant heat leaks. But the thermostat can still only monitor its own location in the room. Hot or cold spots can still develop in other parts of the room.

A better solution is to increase the motion of the air. Air motion will tend to stabilize the temperature. Warmer air can be moved into cold spots, or vice versa. A fan is a simple device for stimulating air motion. The circuit shown in Fig. 5-11 will control a fan in response to ambient temperature. A special purpose IC (LM3911) is used to sense the temperature. The parts list for this project is given in Table 5-5.

The relay is chosen so that its contacts will safely carry the current drawn by the fan. Similarly, the value of fuse F2 should be selected specifically for the fan to be used with the system. Typically, this fuse should not be greater than 1 or 2 A at the very most. Do not, under any circumstance, omit either of the fuses shown in the diagram.

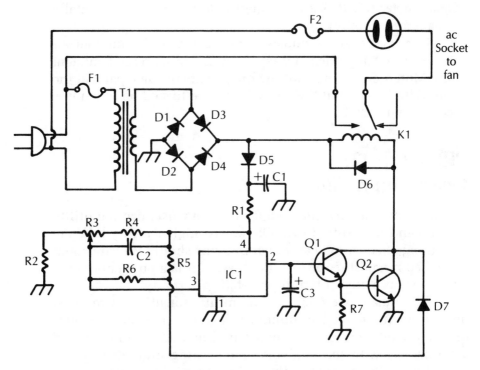

Fig. 5-11   *This temperature equalizer circuit will help you get the maximum benefit out of your heating/cooling system.*

Table 5-5   Parts List for the Temperature Equalizer of Fig. 5-11.

IC1	LM3911 temperature sensor control IC
Q1, Q2	NPN transistor (2N2222 or similar)
D1 – D6	1N4002 diode
D7	1N457 diode
K1	24-V relay (contacts current handling should be selected to suit driven fan)
T1	24-V, 0.5-A transformer
F1	0.5-A fuse (3AG)
F2	Fuse (3AG) (value determined by driven fan)
C1	100-μF, 100-V electrolytic capacitor
C2	0.1-μF, 35-V capacitor
C3	5-μF, 25-V electrolytic capacitor
R1	12-kΩ resistor
R2, R4	27-kΩ resistor
R3	5-kΩ potentiometer
R5	100-kΩ resistor
R6	10-MΩ resistor
R7	22-kΩ resistor

The temperature that switches on the fan is determined by potentiometer R3. Its setting is best determined experimentally for maximum comfort.

Of course, a continuously running fan will stimulate air movement, but at a price of wasted power and increased ambient noise. Moreover, it can result in chilly drafts in some parts of the room. With this circuit the fan will only be turned on when it is specifically needed.

PROJECT 22:
Heater humidifier

In winter months, heaters are put into heavy use, driving utility bills up and humidity down. Obviously, higher utility bills are something of concern to all of us. But why should we care about decreased humidity? For one thing, arid air is a poor conductor, which means a greater buildup of static electricity.

The human body is also humidity sensitive. The most healthy environment has humidity levels in the 30% to 50% range. In winter months, the house is sealed off, there is limited air exchange with the outdoors, and heating units dry up the humidity in the air. Often the indoor humidity in winter can drop to 10% to 20%, which is extremely dry. This can irritate sensitive membranes, leading to sore throats and other such ailments. In addition, the body's humidity sensitivity can make dry air in the 50° to 70°F range seem even colder than normal. Raising the humidity to 30% to 50% will permit a lower thermostat setting for a given degree of comfort.

This project is designed to add humidity to heated air. Before I get to the actual circuitry, let me voice a few warnings. When artificially adding humidity, you shouldn't go overboard and add too much. That can be as bad as too little. Humidity levels greater than about 60% can significantly increase discomfort, and make cold temperatures seem even colder. Also germs tend to thrive in moist air. Some people, having heard about the problems of winter low humidity, add so much artificial humidity that their homes become positively dank. They are doing more harm than good. A little added humidity is good. Too much is bad.

The circuit shown in Fig. 5-12 can add humidity to heated air. The sensor (again, an LM3911 IC) is placed in the air path of the heater's main vent (or vents). When the temperature exceeds a point set by potentiometer R3, a solenoid is activated, opening a

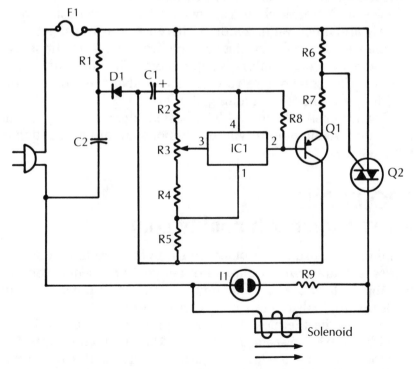

Fig. 5-12 *This circuit adds humidity to dry heated air.*

Table 5-6   Parts List for the Heater Humidifier of Fig. 5-12.

IC1	LM3911 temperature sensor control IC
Q1	PNP transistor (2N3906 or similar)
Q2	Triac (T2302B or similar)
D1	1N2069 diode
I1	NE-2H neon lamp
F1	3-A fuse (3AG)
C1	100-μF, 25-V capacitor
C2	1-μF, 200-V capacitor (nonpolarized)
R1	270-Ω resistor
R2, R4, R9	27-kΩ resistor
R3	10-kΩ potentiometer
R5	4.7-kΩ resistor
R6	1.2-kΩ resistor
R7	1.8-kΩ resistor
R8	150-kΩ resistor
Solenoid	see text
Spray nozzle	see text

spray nozzle. Since the water is sprayed directly into the heating vent, it immediately vaporizes, adding to the air humidity. The parts list for this project is given in Table 5-6.

The spray nozzle should spray a relatively fine mist of water, for easy and quick vaporization. Otherwise, you'll just get a wet heating vent. Suitable nozzles can be purchased from almost any large hardware or plumbing supply store. They are fairly inexpensive. The solenoid valve can be cannibalized from an old washing machine (or something similar).

Potentiometer R3 should be adjusted experimentally for the most comfortable humidity level. Do not set it either too high or too low.

PROJECT 23:
Air conditioner humidity control

Humidity can also be a problem in summer. Here the problem is excess humidity, which makes hot temperatures seem even hotter. You feel sweaty and clammy. An air conditioner not only cools the air, it also drains off excess humidity.

Setting the thermostat can be a problem. In relatively low humidity, 80°F can seem cooler than 77°F in high humidity. Setting the thermostat too low is a waste of energy and may result in too much cooling at times.

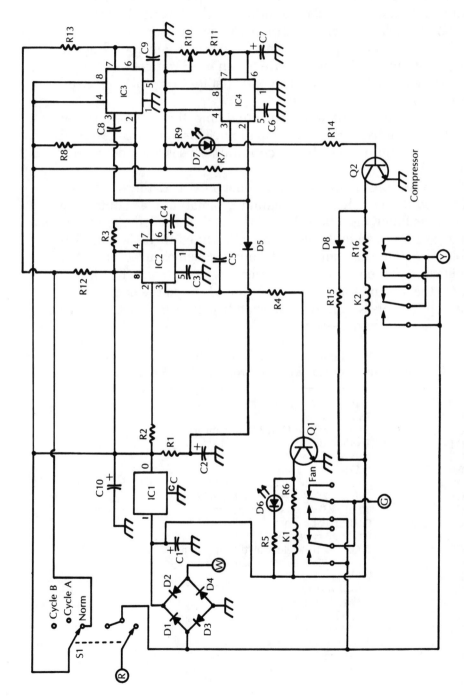

Fig. 5-13 *Summer humidity problems can be minimized with this circuit.*

The problem is, the body responds to a combination of temperature and humidity. A thermostat responds to temperature alone. It does not sense the humidity at all. The circuit shown in Fig. 5-13 cycles the air conditioner's compressor to minimize humidity and air stagnation. This keeps the room cool, but the air conditioner will actually have to run less. You get more comfort while using less power. The parts list for this project is given in Table 5-7.

Since this circuit must be wired into the air conditioner itself, the unit used must be compatible. This circuit is designed on the assumption that the air conditioner uses 24-V dc control circuitry.

Connections to the air conditioner's existing thermostat are indicated on the schematic diagram by small circles containing one of the following letters: G, R, W, and Y. These letters indicate a more or less standard color coding, outlined in Table 5-8. This table also indicates common labeling schemes used in many commercial air conditioner thermostats. The green wire should run to the fan and the yellow wire should run to the compressor. Appropriate relays should be selected to match the current drawn by the fan and compressor.

Table 5-7   Parts List for the
Air Conditioner Humidity Control Circuit of Fig. 5-13.

IC1	7805 + 5-V regulator IC
IC2, IC3, IC4	7555 timer IC
Q1, Q2	NPN transistor (2N2222 or similar)
D1 – D4	1N4004 diode
D5	1N914 diode
D6, D7, D8	LED
K1, K2	DPDT relay (contacts selected to match controlled circuit)
C1	250-μF, 25-V electrolytic capacitor
C2, C4	10-μF, 25-V tantalum capacitor
C3, C6, C9	0.01-μF capacitor
C5, C8	0.001-μF capacitor
C7	1000-μF, 10-V electrolytic capacitor
C10	500-μF, 10-V electrolytic capacitor
R1, R7, R8	27-kΩ resistor
R2	33-kΩ resistor
R3, R12, R13	10-MΩ resistor
R4, R14	4.7-kΩ resistor
R5, R15	3.3-kΩ resistor
R6, R16	1-kΩ resistor
R9	470-Ω resistor
R10	1-MΩ potentiometer
R11	220-kΩ resistor

**Table 5-8   Thermostat Connections
for the Air Conditioner Humidity Control Circuit of Fig. 5-13.**

Schematic label	Cable color	Typical thermostat labels				
G	Green	G	G	G	F	G
R	Red	R	RH	4	M	R5
W	White	W	W	W	H	4
Y	Yellow	Y	Y	Y	C	Y6

A three-position switch (S1)(DP3T) is used to determine the operating mode of the unit. In the NORM position, the air conditioner operates in its normal manner.

In cycle A, potentiometer R10 is used to adjust the compressor on-time. With the component values give in Table 5-7, this time can be set between about 5 minutes to approximately 20 minutes. Then the compressor is shut off for about 20 minutes. The fan will run for about 2 minutes after the compressor is shut down.

Cycle B is very similar to cycle A, but the compressor off time is twice as long. This mode is good for keeping the home reasonably cool and dry when no one is at home.

$\diamond \diamond$ 6

Liquid control

IN THIS CHAPTER WE WILL LOOK AT A NUMBER OF CIRCUITS FOR dealing with liquids. Some of these might not be applications you are likely to immediately consider as candidates for electronic circuitry. The liquids should be kept out of the circuitry itself. Only the probe should come in contact with the liquids.

Liquid probes

Many liquids, including water, are electrically conductive. Liquids are usually more conductive than air. A pair of metallic probes (stiff wires or metal rods) are placed where the liquid is expected, as illustrated in Fig. 6-1. If the liquid level comes up over the ends of the probes, they are electrically shorted together. The equivalent circuit is shown in Fig. 6-2. Essentially it behaves like a switch. When the probes are dry, the switch is open. Immersing the probes closes the switch.

PROJECT 24:
Plant monitor

A variation on the basic liquid level sensing probe can remind you when to water your plants. The circuit is shown in Fig. 6-3. The parts list is given in Table 6-1.

The probes are buried in the soil near the roots. The amount of moisture in the soil determines its conductivity (resistance). A

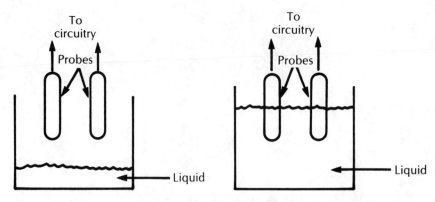

Fig. 6-1 *A pair of metallic probes can serve as a liquid detector.*

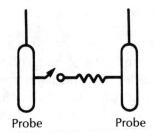

Fig. 6-2 *Liquid probes work because of the electrical conductivity of water and also many other liquids.*

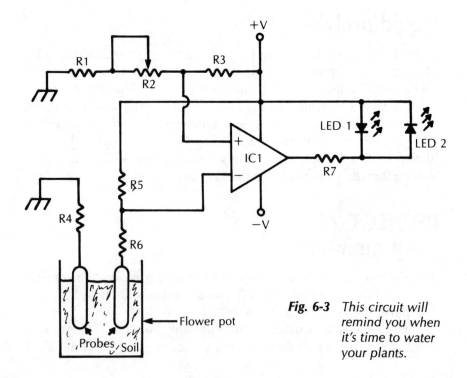

Fig. 6-3 *This circuit will remind you when it's time to water your plants.*

Table 6-1   Parts List for the Plant Monitor of Fig. 6-3.

IC1	Op amp (LM741 or similar)
LED1, LED2	LED
R1, R3, R4, R5	100-kΩ resistor
R2	1-MΩ potentiometer
R6	1-kΩ resistor
R7	330-Ω resistor

comparator determines when the resistance exceeds a level set via the potentiometer. When the soil is sufficiently moist, one LED will be lit. If it gets too dry, the other LED will light up. For best visibility, two different color LEDs should be used.

Alternatively, the comparator could trigger an alarm or buzzer when the soil is too dry. The potentiometer setting should be determined experimentally. When the soil is almost but not quite too dry, adjust the potentiometer until the too dry LED comes on, then back off until the other LED is lit.

PROJECT 25:
Flood alarm

A simple flood alarm circuit is shown in Fig. 6-4. The parts list is given in Table 6-2. When the probes are immersed in water or

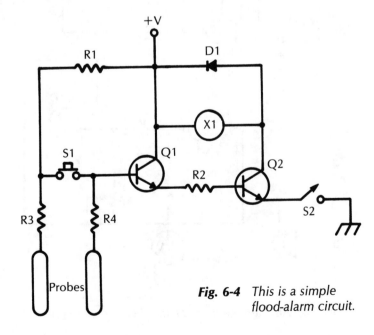

Fig. 6-4 *This is a simple flood-alarm circuit.*

Table 6-2   Parts List for the Flood Alarm of Fig. 6-4.

Q1, Q2	NPN transistor (2N2222 or similar)
D1	1N4002 diode
S1	NO SPST push switch
S2	SPST switch
X1	Alarm device (see text)
R1	33-kΩ resistor
R2 – R4	1-kΩ resistor

another liquid, the alarm will sound. Potentiometer R5 determines the frequency of the alarm tone. A fixed resistance may be substituted, if you prefer.

Place the probes wherever you want to guard against flooding. If water comes up over the ends of the probes, the alarm will sound until the water level goes down.

This circuit is quite flexible. Even the power supply requirements are noncritical. The supply voltage can be anything between +6 and +12 V.

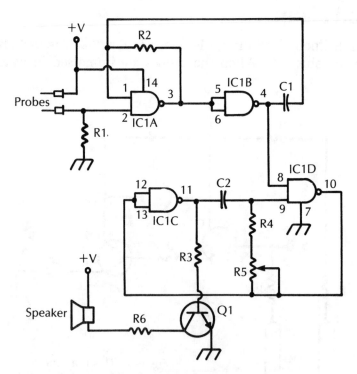

Fig. 6-5 *An alternate flood-alarm circuit is shown here.*

PROJECT 26:
Another flood alarm

An alternative flood alarm circuit is shown in Fig. 6-5. The parts list is given in Table 6-3. It is similar to the previous circuit. Depressing switch S1 permits you to test the alarm. With switch S2 closed, the alarm cannot operate. This allows it to be selectively disabled, as desired.

Table 6-3   Parts List for the Flood Alarm of Fig. 6-5.

IC1	CD4011
Q1	2N3904 transistor
SPKR	Small speaker
C1	0.1-μF capacitor
C2	0.01-μF capacitor
R1	1-MΩ resistor
R2	3.3-MΩ resistor
R3, R4	10-kΩ resistor
R5	50-kΩ potentiometer
R6	33-Ω resistor

The alarm itself can be almost anything. You can use a bell, buzzer, or Sonalert™. Any sound source that can be activated by an applied voltage will do. If you prefer, a visual indicator such as a light can be used in place of the audible alarm.

PROJECT 27:
Moisture detector

The circuits shown in Figs. 6-4 and 6-5 are intended to measure an accumulation of water (or other liquid). A depth of at least one-half inch is required. The probes can be placed higher for greater depths, but they can't really be placed much lower. Suppose we need to detect a smaller amount of liquid?

The moisture detector circuit of Fig. 6-6 should do the trick. The parts list for this project is given in Table 6-4. The sensor is made up of fine wires spaced an inch or so apart. The alarm can be almost anything, as in the last project. A buzzer, bell, or Sonalert™ can be triggered by this circuit. Nothing is terribly critical here. Almost any PNP transistor and SCR can be used with good results.

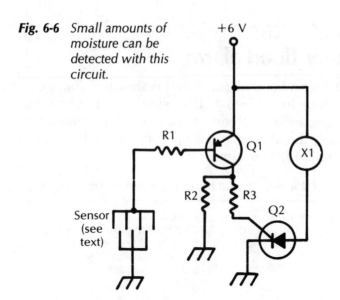

Fig. 6-6 *Small amounts of moisture can be detected with this circuit.*

Table 6-4   **Parts List for the Moisture Detector of Fig. 6-6.**

Q1	PNP transistor (2N3906 or similar)
Q2	SCR
X1	Alarm sounder (buzzer, bell, or Sonalert™)
R1	100-kΩ resistor
R2	10-kΩ resistor
R3	1-kΩ resistor

PROJECT 28:
Sump pump controller

One of the primary applications for a flood detector is to control a sump pump to automatically correct the problem. This is unquestionably automation at work. An automated sump pump controller circuit is illustrated in Fig. 6-7. The parts list for this project is given in Table 6-5.

Notice that this circuit has three probes. The off probe should be mounted somewhat lower than the on probe. The common probe should be lined up with the off probe, or a little lower.

If the water rises enough to touch the common and off probes, nothing happens. However, when the water reaches the on probe, the relays are triggered and the sump pump is turned on.

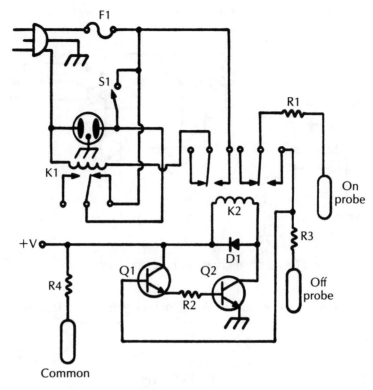

Fig. 6-7 *When flooding is detected, this circuit will turn on a sump pump.*

Table 6-5   Parts List for the Sump Pump Controller of Fig. 6-7.

Q1, Q2	NPN transistor (2N2222 or similar)
D1	1N4002 diode
K1	120-V ac SPST relay (contacts determined by load pump)
K2	DPDT dc relay
F1	Fuse (selected to match load pump)
S1	SPST switch
R1 – R4	1-kΩ resistor

The pump will presumably lower the water level. When the level drops below the on probe, nothing will happen. The pump will keep running. It is not until the water level drops below the off probe that the system will shut down. This will prevent the pump from oscillating on and off with the water level just at the edge of the on probe.

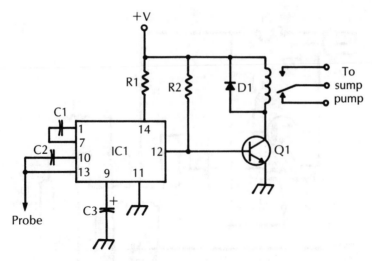

Fig. 6-8 *This is an alternate sump pump controller circuit.*

Table 6-6 Parts List for the Sump Pump Controller of Fig. 6-8.

IC1	LM1830
Q1	2N3055
D1	1N4002
C1	0.001-μF capacitor
C2	0.05-μF capacitor
C3	25-μF capacitor
R1	470-Ω resistor
R2	2.2-kΩ resistor

PROJECT 29:
Another sump pump controller

A different sump pump controller circuit is shown in Fig. 6-8.
The parts list is given in Table 6-6. This circuit is a little simpler
than the previous project because it is built around a single IC
(the LM1830). This circuit has only a single probe and may tend
to oscillate under some circumstances.

Stereo and TV projects

HOME ENTERTAINMENT SYSTEMS OFFER MANY OPPORTUNITIES for remote control and automation applications. Many commercial pieces of audio and video equipment come with remote controllers. This chapter presents several relatively simple projects for increasing your enjoyment of stereo, television, and video equipment.

PROJECT 30:
Stereo automatic shutoff

Most automatic record changers feature automatic shutoff after the last record has been played. This is handy, but power continues to be applied to the rest of the system. The amplifier, for example, remains on until you manually turn it off. It's easy to forget it when no sound is coming out.

If you are recording your records, the tape will keep running after the last record is over. Rewinding back to the end of the last selection when you don't know where it is can be an irritating nuisance.

A simple solution is illustrated in Fig. 7-1. Simply wire an ac socket in parallel with the turntable's motor. When the motor shuts itself off, power will be disconnected from this socket. Whatever is plugged into this socket will be shut down along with the record changer.

Some tape machines have problems if power is interrupted while the tape is running. Such machines should not be used

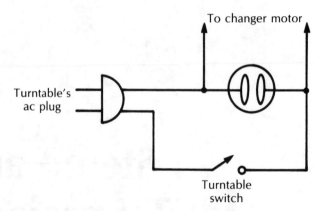

Fig. 7-1 *This circuit will shut off a tape recorder after the last record has been played on an automatic turntable.*

with an auto shutoff system like this. Most modern tape decks, however, have no problem at all, although the tape might need to be slackened slightly by hand before power is reapplied to prevent tape breakage.

While no fuse is shown in the diagram, adding one certainly wouldn't be a bad idea. The fuse's rating will depend on what is being plugged into the socket. Even if each piece of equipment in your stereo system is independently fused, additional fusing isn't necessarily overkill. You can never have too much overcurrent protection.

PROJECT 31:
Advanced auto shutoff

A more deluxe automatic shutoff circuit is shown in Fig. 7-2. Once again, the circuit is placed in parallel with the record changer's motor. When the record changer turns itself off, the DPDT relay is triggered, removing power from the ac sockets.

Switch S1 is a simple DPDT switch that allows you to set the system into the automated mode or the manual mode. In the manual mode, the record changer does not control power to the sockets. This allows maximum flexibility. You don't want to have to run the record player or rewire the system to listen to the FM radio or a prerecorded tape.

Multiple ac sockets can be wired in parallel. Fusing is advised although it is not shown here. The two lamps (I1 and I2)

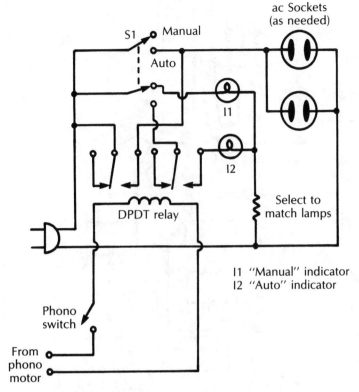

ac Sockets
(as needed)

S1 Manual

Auto

I1

I2

DPDT relay

Select to
match lamps

I1 "Manual" indicator
I2 "Auto" indicator

Phono
switch

From
phono
motor

Fig. 7-2 *This is a somewhat more sophisticated automatic shutoff circuit for
stereo systems.*

and the resistor are optional. These are simply indicators of S1's
position. In the manual mode, I1 is lit; I2 lights up for the auto
mode.

Despite the extreme simplicity of this project (there isn't
even a parts list), it can be a very powerful and valuable addition
to almost any stereo system.

PROJECT 32:
VOX recorder controller

Do you ever do any dictation into a tape recorder? If so, and if
you're like most of us, there are probably a lot of pauses on the
tape as you stop to think. This wastes tape and can slow tran-
scription. But turning the recorder on and off manually can be a
nuisance and a distraction. Wouldn't it be nice if the tape re-

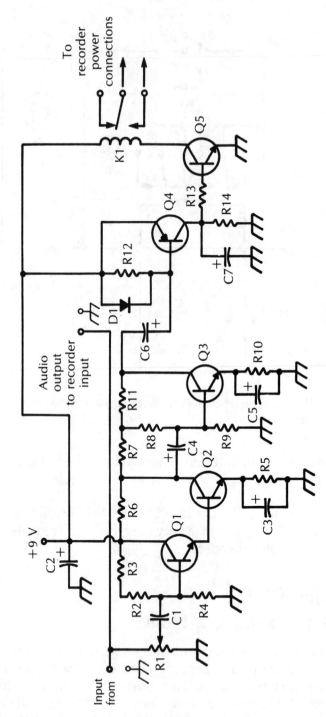

Fig. 7-3 This VOX recorder controller will avoid wasting tape caused by recording unwanted pauses.

corder was smart enough to turn itself on and off at appropriate times?

Well, while I'm not sure it really counts as intelligence on the part of your recorder, the circuit shown in Fig. 7-3 will turn on the recorder whenever you speak into the microphone. When you stop speaking, the system will wait a few seconds then shut off the recorder. This eliminates any long pauses on the tape. The parts list for this circuit is given in Table 7-1.

Table 7-1 Parts List for the VOX Recorder Controller of Fig. 7-3.

Q1, Q2, Q3, Q5	NPN transistor (2N4256 or similar)
Q4	PNP transistor (2N428 or similar)
D1	1N38B diode
C1	0.1-μF capacitor
C2, C3, C5	100-μF, 50-V electrolytic capacitor
C4, C6, C7	10-μF, 50-V electrolytic capacitor
R1	2-MΩ potentiometer
R2	1-MΩ resistor
R3	100-kΩ resistor
R4	270-kΩ resistor
R5, R10	560-Ω resistor
R6, R9, R11	2.7-kΩ resistor
R7	1.2-kΩ resistor
R8	15-kΩ resistor
R12	33-kΩ resistor
R13	47-kΩ resistor
R14	22-kΩ resistor
K1	dc relay (see text)

The recorder is controlled by relay K1. The exact connections and relay required will depend on the specific tape machine you use. This circuit is very easy to use. There is just a single control. Potentiometer R1 controls the sensitivity of the input stage to allow for any background noise in the area. Simply adjust this control so that it activates the recorder when you speak and shuts it off when you are not speaking into the microphone.

This circuit can easily be adapted to control almost anything by sound, such as shouting or clapping your hands. The next project is specifically designed for such applications.

PROJECT 33:
VOX relay

A general-purpose VOX (voice-operated switch) circuit is shown in Fig. 7-4. The parts list is given in Table 7-2. Almost any small relay can be activated with this circuit, permitting it to control almost anything. If a larger load is used, the small relay can be used to control a second, larger relay.

Potentiometer R1 controls the sensitivity of the circuit. This means you can adjust how loud the triggering sound must be. Any loud, sharp sound can be used to activate the VOX circuitry. A good, sharp hand clap will usually give good results. The sensitivity control should be set high enough that false triggering from background noise won't be a major problem.

The 555 timer introduces a slight delay. By adjusting potentiometer R4, delays from about 0.05 second to approximately 5 seconds can be set up. For longer delays, a louder sound may be required. At the higher settings of the delay control, a hand clap will be too short. A shout or whistle will work better under such

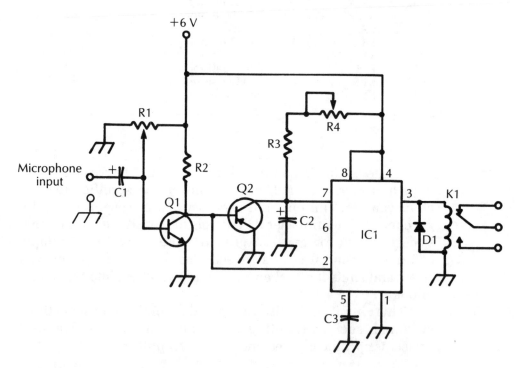

Fig. 7-4 *Almost anything can be switched by a sound with this VOX relay circuit.*

Table 7-2   Parts List for the VOX Relay of Fig. 7-4.

Q1	NPN transistor (2N3904 or similar)
Q2	PNP transistor (2N3906 or similar)
IC1	555 timer IC
D1	1N4002 diode
K1	Relay to suit controlled device
C1	1-μF, 10-V electrolytic capacitor
C2	50-μF, 10-V electrolytic capacitor
C3	0.01-μF capacitor
R1	10-kΩ potentiometer
R2, R3	1-kΩ resistor
R4	100-kΩ potentiometer

circumstances. The advantage of the longer time delay is that it reduces the chances of some stray transient sound accidentally activating the relay at an undesired time.

PROJECT 34:
Sound compressor

Have you ever tried recording a panel discussion or some other event with several people talking from various locations? If you have, you're certainly familiar with the hassles of constantly watching the VU meters and riding the tape recorder's gain controls. There is always at least one person who constantly bellows, and another one who barely speaks above a whisper. How can you set the gain controls so that both will be recorded clearly without undue distortion? It can especially be a problem when you don't know who is going to speak next. Sometimes even recording a single speaker can lead to problems of this sort. Many people constantly raise and lower the volume of their voices as they speak, often in an unpredictable manner.

Wouldn't it be nice if you had an assistant with incredibly fast reflexes, and no tendency to be distracted or let his mind wander? You could put this assistant to work riding the gain controls, freeing you for other tasks. As you probably suspect, this project is just such an automated assistant. The official name for this type of circuit is a sound compressor.

The circuit shown in Fig. 7-5 literally compresses the dynamic (volume) range of the signals being recorded. Loud sounds are attenuated to a lower level, and soft sounds are

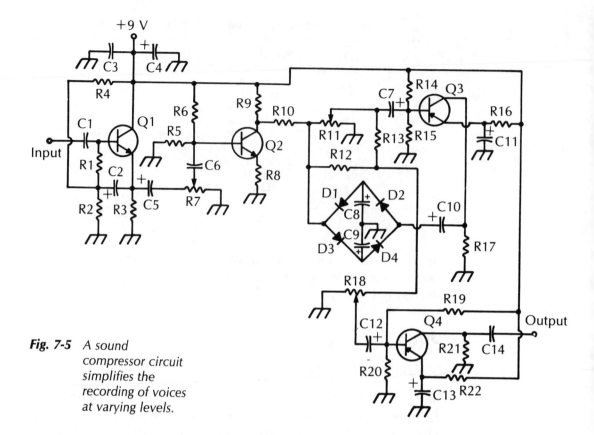

Fig. 7-5 *A sound compressor circuit simplifies the recording of voices at varying levels.*

boosted to a higher level. The parts lists for this project is given in Table 7-3.

The circuit responds to changes in sound level very rapidly. There are three controls: R7, R11, and R18. Potentiometer R7 adjusts the input sensitivity. This control determines the softest sound level that can be recognized by the circuitry.

The average output level to the tape deck is set by potentiometer R18. The output signal will be relatively constant, within a narrow range, and the recorder's gain controls can be set to a good position and then left alone.

The third control, potentiometer R11, sets the amount of compression. This is the dynamic range control, determining the amount of variation in the output signal to be recorded.

A compression circuit is great for equalizing voice levels, but it should not be used to record music. Music generally depends heavily on dynamic variation for its emotional effect. Compressed music sounds very flat, dull, and extremely uninteresting.

Table 7-3   Parts List for the Sound Compressor Circuit of Fig. 7-5.

Q1, Q2	NPN transistor (2N2923 or similar)
Q3, Q4	PNP transistor (2N397 or similar)
D1, D2	1N914 diode
D3, D4	1N270 diode
C1, C14	0.5-μF capacitor
C2, C7, C12	33-μF, 25-V electrolytic capacitor
C3	0.022-μF capacitor
C4	150-μF, 25-V electrolytic capacitor
C5, C6	4.7-μF, 25-V electrolytic capacitor
C8, C9, C11, C13	47-μF, 25-V electrolytic capacitor
C10	100-μF, 25-V electrolytic capacitor
R1, R2, R15	56-kΩ resistor
R3, R9	2.2-kΩ resistor
R4, R14	22-kΩ resistor
R5, R6	15-kΩ resistor
R7	25-kΩ potentiometer
R8	120-Ω resistor
R10, R21	3.3-kΩ resistor
R11, R18	100-kΩ potentiometer
R12	8.2-kΩ resistor
R13, R17, R22	1.2-kΩ resistor
R16	560-Ω resistor
R19	33-kΩ resistor
R20	120-kΩ resistor

This is an example of using the right tool for the right job. You can't expect a simple circuit to handle every job, of course.

PROJECT 35:
Recorder timer switch

Most VCRs (video cassette recorders) include a timer so programs can be recorded unattended. But suppose there is a program coming on the radio that you'd like to record, but you're not going to be home.

You can get a mechanical ac timer, but you probably won't be very satisfied with the results. These mechanical timers are intended primarily for lighting. High accuracy in the timing is not particularly critical and is definitely not a feature of these devices. They can be off by as much as 5 to 15 minutes. For automatic lighting applications, this degree of error is probably acceptable. For recording, the error is quite unacceptable. Either you'll miss part of the program (when the timer runs slow), or

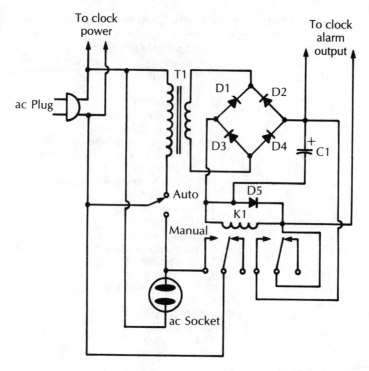

Fig. 7-6   *An electronic alarm clock can serve as a recorder timer.*

you'll record a lot of stuff you're not interested in before the show (when the timer runs fast).

Far better results can be achieved by using a true time-keeping clock, especially an all-electronic (digital) clock with an alarm feature. The circuitry is shown in Fig. 7-6, and the parts list is given in Table 7-4. As you can see, this circuit is not very complicated, and requires only a handful of parts. The existing clock does most of the work.

When the alarm goes off, the relay is activated. The relay is held in for 59 minutes. When the switch is in the automatic position, power is fed through the ac socket while the relay is activated.

Table 7-4   Parts List for the Recorder Timer Switch of Fig. 7-6.

T1	12-V ac transformer
D1 – D5	1N4002 diode
C1	100-μF, 25-V electrolytic capacitor
K1	12-V dc DPDT relay

The automatic timer function can be bypassed by placing the switch in the manual position. When the switch is in this position, power is applied directly to the ac socket. The clock alarm and relay are ignored.

PROJECT 36:
TV auto-off circuit

Do you ever fall asleep with the TV on and wake up in the wee hours of the morning with a snowy, no-signal screen? Or worse, you're rudely awakened when the station signs back on the air—usually much earlier than you want to get up. This is annoying, of course. It also wastes power. Large screen color TVs eat up quite a bit of power.

The circuit shown in Fig. 7-7 will automatically disconnect power when the station stops broadcasting. The 555 timer provides a slight delay before shutoff so any momentary signal blanking between programs (common when a station switches between local and network programming) won't turn your TV off. The signal must be absent for a minute or so. The exact delay can be set via potentiometer R2. A bypass switch is provided so that the TV set can be used normally if desired. Once the system has shut down, the set will stay off until the reset button is depressed. The parts list for this project is given in Table 7-5.

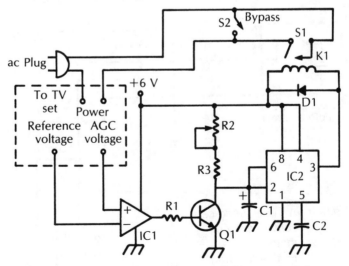

Fig. 7-7 *This circuit will automatically turn off the TV when the station goes off the air.*

Table 7-5   Parts List for the TV Auto-Off Circuit of Fig. 7-7.

IC1	LM339 comparator
IC2	555 timer
Q1	NPN transistor (2N2222 or similar)
D1	1N4002 diode
K1	dc Relay with SPST contacts
S1	NO momentary contact SPST switch
S2	SPST switch
C1	10-μF, 25-V electrolytic capacitor
C2	0.01-μF capacitor
R1	1-kΩ resistor
R2	1-MΩ potentiometer
R3	390-kΩ resistor

This circuit can also be adapted for use with your stereo system, so you can doze off while listening to the radio, a tape, or a record. Naturally, if you're tuned to a 24-hour station, this project won't do much of anything. But for stations that shut down between midnight and dawn (or thereabouts), this device can be handy.

PROJECT 37:
Remote control mute

Suppose you're watching TV or listening to the stereo when the phone or the doorbell rings. Often it would be convenient to be

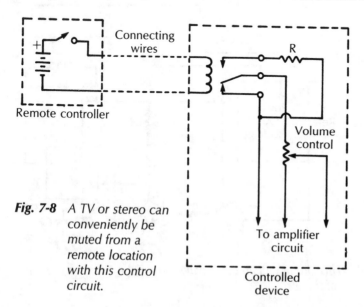

Fig. 7-8   *A TV or stereo can conveniently be muted from a remote location with this control circuit.*

able to lower the volume by remote control. Figure 7-8 illustrates how this can be done. When activated, an extra resistance is switched into the circuit in series with the existing volume control potentiometer.

The amount of attenuation will depend on the value of the added resistor. As a rule of thumb, I'd say the resistor value should be about one-third to one-half the value of the potentiometer. For example, if the volume control is a 1-MΩ potentiometer, use a 470-kΩ resistor for the added resistance.

❖ 8
Telephone projects

THE TELEPHONE IS A NATURAL CHOICE FOR REMOTE CONTROL AND automation projects. In fact, the telephone itself is a remote control device of sorts. The very name suggest this. The word "telephone" can be freely translated as "distance-sound." Think about what happens when you call someone. You dial a number at one location, causing a string of relays to close at another location (the telephone company's trunk lines), causing another phone to ring at still another location.

In this chapter we will carry the remote control aspects of the telephone even further. It is a very handy control device. Most of the projects in this chapter are designed to connect directly to the phone lines. Legally you are obligated to contact the local phone company before you connect anything to their lines. It is possible to disrupt phone service to hundreds of other people or damage expensive phone company equipment by tapping off the phone lines. None of these projects should cause any problems, but the phone company has the right to know what you are plugging into their lines.

PROJECT 38:
Telephone-activated relay

The circuit shown in Fig. 8-1 will activate a relay when the phone rings. The relay used will depend on the device being controlled. If you have a separate unlisted number just for this purpose, you can control almost anything simply by phoning

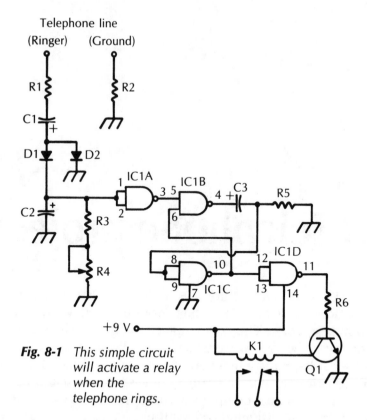

Fig. 8-1 *This simple circuit will activate a relay when the telephone rings.*

home. That might be a bit extravagant for most of us. A more practical application would be for the relay to turn on a light when the telephone rings. This could be useful in a noisy environment or for the hearing impaired. Another useful application would be for the relay to activate an oscillator or other sound generating device to serve as a remote ringer. This is useful when you are outside or if you have a large house. The parts list for this project is given in Table 8-1.

Table 8-1. Parts List for the Telephone-Activated Relay of Fig. 8-1.

IC1	CD4011 quad NAND gate
Q1	NPN transistor (2N2222 or similar)
D1, D2	1N4002 diode
K1	Relay selected for desired application
C1, C2	1-μF, 50-V electrolytic capacitor
C3	100-μF, 50-V electrolytic capacitor
R1, R2	390-kΩ resistor
R3, R5	1-MΩ resistor
R4	1-MΩ potentiometer
R6	1-kΩ resistor

PROJECT 39:
Improved telephone-activated relay

The project described in the last section is certainly functional, but it has definite limitations for true remote control applications. For one thing, it must be connected directly to the telephone lines. This means that you are legally required to get permission from the local telephone company. At best, this is a time-wasting nuisance.

The major limitation of the preceding project is that it is subject to false activation whenever anyone else calls your number. You could set up a separate line just for control purposes but this might be unduly expensive. Besides, it is no guarantee that no one else will ever ring the number. Even unlisted numbers get wrong number calls. Many commercial telephone solicitation outfits use computerized dialing systems that randomly dial valid number combinations, whether they are listed or not, so they can force their sales pitches on everyone, even those who have gotten unlisted numbers. (If I might editorialize for a moment—telephone solicitors are notoriously rude.)

All in all, the circuit of Fig. 8-1 probably isn't very practical for most serious remote control applications. It was presented here primarily as an introduction to this project. The circuit for the improved telephone-activated relay is shown in Fig. 8-2. The parts list is given in Table 8-2.

No direct electrical connection is made to the telephone or its wiring. The ringing is picked up by a pair of crystal microphone elements. The microphones should be placed together near the phone. Why two microphones, instead of just one? This will be explained shortly.

When the phone rings, the sound is picked up by the microphones, triggering SCR Q7. This closes relay K2, activating a simpler timer circuit made up of UJT Q8 and its associated components. The time constant of this timer is determined by resistor R16 and capacitor C8. With the component values given in the parts list, the delay will be about 30 seconds.

If a second ring is picked up by MIC1, SCR Q3 activates relay K1. The circuit shuts down for a period set by the timer built around UJT Q4. The time constant for this timer (determined by resistor R8 and capacitor C4) is a little over 1 minute. If, however, no second ring is detected by the time Q8 times out, the multivibrator made up of transistors Q9 and Q10, along with their asso-

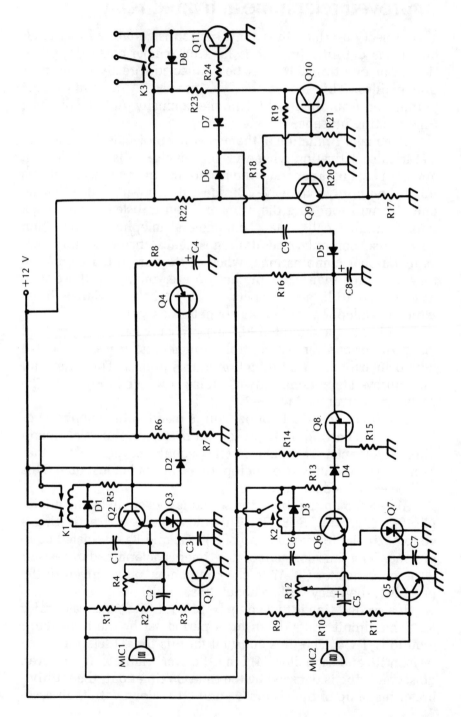

Fig. 8-2   Almost anything can be controlled with this improved telephone–activated relay circuit.

**Table 8-2   Parts List for the
Improved Telephone-Activated Relay of Fig. 8-2.**

Q1, Q2, Q5, Q6, Q9, Q10, Q11	NPN transistor (2N3904 or similar)
Q3, Q7	SCR (C106Y or similar)
Q4, Q8	UJT (2N2646 or similar)
D1, D3, D8	1N4002 diode
D2, D4, D5, D6, D7	1N914 diode
K1	SPDT 12-V dc relay
K2	SPST 12-V dc relay
K3	12-V dc relay (contacts to suit application)
C1, C6, C9	0.1-μF capacitor
C2, C5	30-μF, 25-V electrolytic capacitor
C3, C7	0.01-μF capacitor
C4, C8	100-μF, 25-V electrolytic capacitor
R1, R9, R20, R21	10-kΩ resistor
R2, R10	22-kΩ resistor
R3, R11	100-kΩ resistor
R4, R12	5-kΩ potentiometer
R5, R13, R24	1-kΩ resistor
R6, R14, R17	470-Ω resistor
R7, R15	100-Ω resistor
R8	680-kΩ resistor
R16	27-kΩ resistor
R18, R19	33-kΩ resistor
R22, R23	3.9-kΩ resistor

ciated components, is triggered, activating relay K3, which is connected to the controlled device.

Note that the relay will be activated, turning on the controlled device, if the telephone rings once and only once. To use the remote control system, just dial your own number, let it ring once and hang up. Anyone else calling you will probably let the phone ring at least two or three times, so the relay will not be activated by their calls. False triggering is still possible, but it is rather unlikely.

Once the relay is activated, it stays activated, until another single ring is detected, retriggering the Q9/Q10 multivibrator and deactivating the relay. This project can be put to work in a wide variety of practical remote control applications.

PROJECT 40:
Off-hook alarm

Many families have several extension phones, all on the same number. It can be disturbing and embarrassing to pick up the

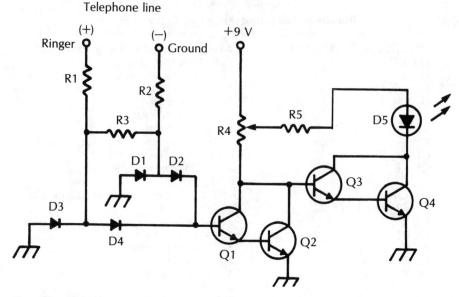

Fig. 8-3 *This circuit will alert you whenever an extension phone is off the hook.*

phone to make a call and interrupt another conversation. Or, perhaps you'd like to know when the kids are using the phone.

This project will help keep the battle of the extension phone under control. The circuit shown in Fig. 8-3 causes the LED to light up when the phone is off the hook. While this feature is probably of limited value, the LED also indicates when the phone is being dialed by blinking on and off. The LED also blinks when the phone rings.

The parts list for this simple project is shown in Table 8-3. If the LED is part of an optoisolator, the off-hook alarm can control an audible alarm or almost anything you choose.

Table 8-3   Parts List for the Off-Hook Alarm of Fig. 8-3.

Q1-Q4	NPN transistor (2N3904 or similar)
D1-D4	1N914 diode
D5	LED
R1, R2, R4	10-MΩ resistor
R3	1.2-MΩ resistor
R5	1-kΩ resistor

PROJECT 41:
Telephone recorder controller

Some people might need a record of their phone calls. The circuit shown in Fig. 8-4 automatically turns on a cassette recorder whenever the phone is off the hook, recording all conversations.

Most standard cassette tape recorders have jacks for an external microphone and a remote switch. This project uses both of these jacks. If the jacks are not available, the recorder must be adapted (jacks added) to be used with this circuit.

When the telephone handset is taken off the hook, the two transistors close the remote switch contacts, activating the tape recorder. The audio signals through the handset are coupled through a capacitor to the microphone input of the tape recorder. By leaving this system set up, all calls can be recorded.

It is a good idea to inform the other party that the conversation is being recorded. It is not legal or moral to use these (or similar) devices to "bug" a phone and record someone else's calls

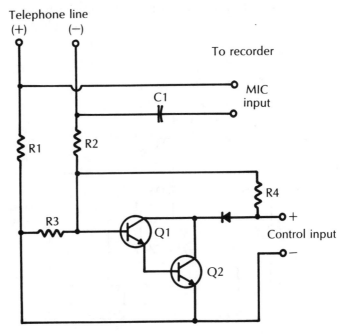

Fig. 8-4 *You can automatically record your telephone conversations with this circuit.*

Table 8-4   Parts List for the Automatic Telephone Recorder of Fig. 8-4.

Q1, Q2	NPN transistor (2N4954 or similar)
C1	0.022-μF capacitor
R1	1.2-kΩ resistor
R2	270-kΩ resistor
R3	68-kΩ resistor
R4	33-kΩ resistor

without their knowledge. This project is intended for personal use only. The parts list for this circuit is given in Table 8-4.

PROJECT 42:
Autodialer

Automatic dialers are popular telephone accessories. Many commercial autodialers have been placed on the market. They are convenient, but they are fairly expensive.

A "quick and dirty" autodialer circuit is shown in Fig. 8-5. The parts list for this project is given in Table 8-5. This circuit

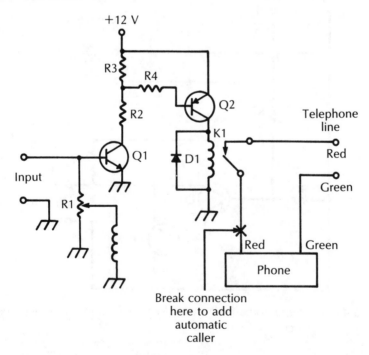

Fig. 8-5 *Automatic telephone dialers are popular accessories.*

Table 8-5   Parts List for the Autodialer of Fig. 8-5.

Q1	NPN transistor (2N3904 or similar)
Q2	PNP transistor (2N3906 or similar)
D1	1N4002 diode
K1	12-V dc relay with NC SPST relay
R1	500-kΩ potentiometer
R2	12-kΩ resistor
R3	1-kΩ resistor
R4	2.2-kΩ resistor
L1	Telephone pickup coil (part of the telephone's existing circuitry)

uses the dial-pulse type of dialing signals, so a Touch Tone™ system is not required. Dial pulses can be recorded on a cassette tape recorder. Playing the tape back through the input of this circuit causes the stored number to be automatically dialed.

A more versatile approach is to use a computer to generate the dial pulses. Depending on your adeptness at programming, almost any special features you choose can be incorporated into the system.

A dial telephone works by repeatedly opening and closing the power connection to the phone. The circuit shown here uses a relay to perform this function. It is connected directly to the telephone line. If you prefer to avoid any potential legal hassles, or want a more portable system, a solenoid can be used in place of the relay to mechanically depress and release the phone's cradle button.

PROJECT 43:
Automatic caller

The circuit shown in Fig. 8-6 is intended for use with the preceding project. It is inserted into the earlier circuit by breaking the connection marked with an "X" in Fig. 8-5 and wiring in this new circuitry.

Any type of NC (normally closed) alarm switch can be used for this circuit. A typical application uses intrusion switches for a burglar alarm. When one of the monitored switches is opened, the tape recorder is activated. A previously recorded number is then dialed. A previously recorded message can be played, alerting whoever answers the phone to the alarm condition. Record the spoken message two or three times. You can't predict how long it will take the other party to answer the phone, and you

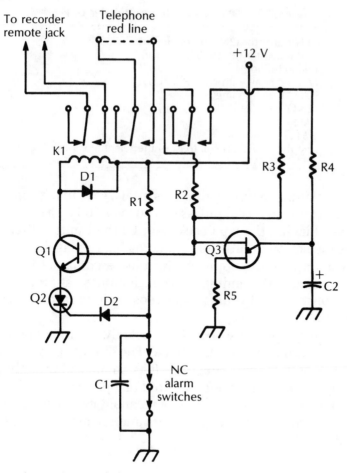

Fig. 8-6 *The automatic dialing circuit of Fig. 8-5 can be adapted to auto-matically call for help during an emergency.*

Table 8-6   Parts List for the Automatic Caller of Fig. 8-6.

Q1	NPN transistor (2N3904 or similar)
Q2	SCR (C106Y or similar)
Q3	UJT (2N2646 or similar)
D1	1N4002 diode
C1	0.1-μF capacitor
C2	330-μF, 25-V electrolytic capacitor
R1	56-kΩ resistor
R2	1-kΩ resistor
R3	470-Ω resistor
R4	68-kΩ resistor
R5	100-Ω resistor
K1	Relay

want to be sure they hear the entire message. The parts list for this project is given in Table 8-6.

PROJECT 44:
Remote telephone ringer

Have you ever missed an important call because you were outside or in another room and didn't hear the telephone ring? If so, you'll certainly welcome this project. It is a simple remote telephone ringer. It can be installed almost anywhere with access to a telephone jack. A clear, hard-to-ignore tone will be heard each time the phone rings. The schematic diagram for this project is shown in Fig. 8-7, and the parts list is given in Table 8-7.

Notice that this circuit is very similar to the one we used in project 38. Instead of controlling a relay driver, in this project, the telephone line sensor circuit drives a tone generator circuit whenever a ring signal is detected on the phone lines.

The tone generator is built around the two timer sections of a 556 dual timer IC. If you prefer, you can substitute two separate 555 single timer ICs. Just correct the pin numbers.

With the component values given in the parts list, a pleasant mid-frequency warble tone will be heard from the remote ringer. If you'd like to experiment with different sounds, try changing the values of resistors R6, R7, and R8, and capacitors C4 and C6. There is little to be gained from experimenting with the other component values in this circuit. Nothing is particularly critical here.

Potentiometer R4 adjusts the control circuit's sensitivity. Set this potentiometer so the remote ringer sounds reliably when the phone rings, but does not give a false signal because of noise on the phone lines. I'd recommend using a screwdriver-adjusted trimpot for this control. If the project does not work properly, try reversing the "ring" and "tip" connections.

Potentiometer R9 is a volume control to adjust the sound level through the small speaker for the remote ringer. In some applications, a volume control won't be needed or wanted. If this is the case in your application, simply replace this potentiometer with an appropriately valued fixed resistor. The smaller this resistance is, the louder the sound heard through the speaker will be. Capacitor C8 protects the speaker's coil from any dc content in the output signal.

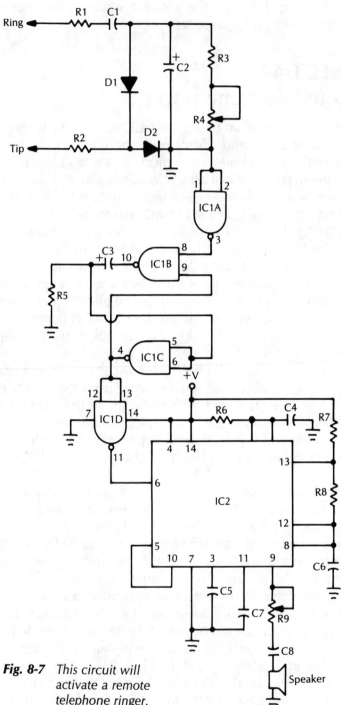

Fig. 8-7 *This circuit will activate a remote telephone ringer.*

Table 8-7   Parts List for the Remote Telephone Ringer Circuit of Fig. 8-7.

IC1	CD4011 quad NAND gate
IC2	556 dual timer (or two 555 timer ICs)
D1, D2	1N4002 diode (or similar)
C1	0.5-μF capacitor
C2	2-μF, 100-V electrolytic capacitor
C3	100-μF, 25-V electrolytic capacitor
C4, C6, C8	0.1-μF capacitor
C5, C7	0.01-μF capacitor
R1, R2, R3	470-kΩ, 0.5-W resistor
R4	1-MΩ potentiometer (sensitivity)
R5	1-MΩ, 0.25-W resistor
R6	330-kΩ, 0.25-W resistor
R7	12-kΩ, 0.25-W resistor
R8	4.7-kΩ, 0.25-W resistor
R9	500-Ω potentiometer (volume)

Chapter summary

The telephone is an ideal candidate for remote control projects, but you must keep the legal aspects in mind. Other people rely on the same telephone lines that you're using. Anything that could possibly disrupt regular services must be avoided at all costs.

Use of the telephone lines is regulated by the FCC (Federal Communications Commission). All customer-supplied equipment must be FCC-type approved to be legally connected to the telephone lines. For these projects (the ones that connect directly to the phone lines) an approved protective coupler must be used to be legal. Check with your local phone company for details.

Don't try to bypass these regulations. Some circuits will alter the signals on the lines enough to be detected by the phone company's monitoring equipment, and you could be subject to serious legal penalties.

❖ 9

Controlling motors

MANY REMOTE CONTROL AND AUTOMATION PROJECTS INVOLVE some sort of mechanized physical motion. In most instances, this requires some sort of motor. Motors were discussed briefly in chapter 1. In this chapter we will explore the use of these devices in a little more detail. Particular emphasis will be placed on practical techniques for controlling motors with various electrical signals. Before getting to the practical applications, however, let's spend a little time getting better acquainted with the motor itself.

There are actually several different types of motors. Some motors are designed to run on a dc voltage, while others require an ac voltage. The dc motors include series, shunt, and compound types. Each type of motor has its own advantages and disadvantages. In most cases, the best choice will depend on the specific application at hand.

dc motors

The dc motor tends to be more commonly used in control applications than ac motors. They tend to be more readily available, smaller, lighter, and less expensive than their ac counterparts.

Basically, a dc motor is made up of a moving coil called the armature winding and a stationary coil called the field winding. A current flows through both coils generating a magnetic field around each coil. The armature rotates, producing a rotating magnetic field. The fixed position of the field winding results in

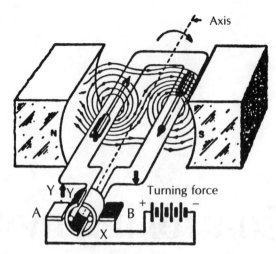

Fig. 9-1 *A dc motor operates from the interaction of two magnetic fields.*

a stationary magnetic field. These two magnetic fields cause the motor's shaft to rotate. The basic structure of a simple dc motor is illustrated in Fig. 9-1. In some motors, a permanent magnet is used in place of the field winding to produce the stationary magnetic field. The effect is the same.

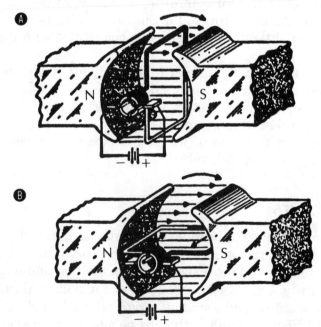

Fig. 9-2 *The armature winding rotates, while the field winding remains stationary.*

Figure 9-2 illustrates the operation of a dc motor. In A, the armature's position causes it to produce a magnetic field with the north pole at the top and the south pole at the bottom. Unlike magnetic poles attract and like magnetic poles repel. This forces the armature to rotate in a counterclockwise direction because of the interaction of the two magnetic fields.

Soon the armature reaches the position shown in B. The unlike poles are now lined up. If that was all there was to the motor, it would stop rotating at this point. Certainly that wouldn't be a very useful device.

The secret to the operation of a dc motor lies in the way current is applied to the armature winding. A brush and commutator arrangement is used. It reverses the polarity of the current connections to the armature winding at the point in its rotation where the torque would drop to zero if no change was made. Now the poles of the armature's magnetic field are reversed, so the two magnetic fields are no longer lined up. Unlike magnetic poles repel each other, and the armature keeps turning. Just as it reaches the point when the two fields would line up, the connections are reversed again, and we're right back where we started.

The armature is forced to keep rotating because the two magnetic fields can never line up in a position of stability. The motor shaft is connected to the armature, so it continuously rotates too. Of course, the motor described here has been greatly simplified, but this brief outline gives you the basic idea of how a dc motor functions.

Most practical dc motors have multiple armature windings, each with its own pair of commutator segments. With just one armature winding, as in our example, the torque would be uneven and the motion jerky. The torque would be highest when the unlike poles were coming close together, and would be significantly lower during the rest of the rotation. Multiple armature windings result in much smoother operation.

The motor we have described has just two magnetic poles from the field and is called a two-pole motor. While two-pole motors do exist, most practical motors have four or more poles.

A dc motor must have current flowing through both the armature and the field windings in order to run. The way in which the two windings are connected to each other has a significant effect on the operating characteristics of the motor, in terms of speed and torque changes with varying load conditions.

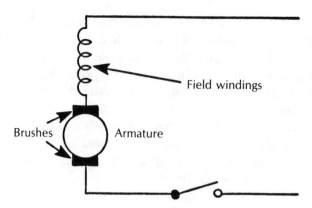

Fig. 9-3 *In a series motor, the field and armature windings are wired in series.*

There are basically three types of dc motors, distinguished by the connections between the windings. They are

- the series motor
- the shunt motor
- the compound motor.

In the series motor, the field and armature windings are wired in series, as illustrated in Fig. 9-3. Because of the series arrangement, all of the armature current must necessarily flow through the field winding too. The field windings of a series motor consists of comparatively few turns of heavy wire.

When power is applied to a series motor with no load, a fairly large current flows through the windings. Torque increases with both armature current and field strength. You can see that this type of motor will have a very high starting torque. The motor's shaft will rotate faster and faster, developing a high counter-EMF (electromotive force). The counter-EMF is a voltage generated in the armature winding by its rotation through the stationary magnetic field. After all, the differences between a motor and a generator are slight.

Returning to our no-load series motor. As the speed increases, the field is weakened, allowing the armature to rotate even faster. A runaway feedback effect occurs. With no load, a series motor will soon destroy or damage itself by rotating at extremely high and ever-increasing speeds. A series motor should never be operated without a load.

If we connect a mechanical load to the motor shaft, the series motor will have to slow down. The greater the load, the lower the speed. As the speed is reduced, the counter-EMF is also reduced, while the current and torque are increased.

To summarize, a series motor has a high starting torque and its speed is determined primarily by the mechanical load. Series motors are usually employed in applications requiring a high torque for a relatively short period of time.

The second type of dc motor is the shunt motor. As illustrated in Fig. 9-4, the armature winding is connected in parallel with the field winding. The field current in a shunt motor is considerably smaller than the field current in a series motor. The field strength can be made as large as required by increasing the number of turns of wire in the field winding. Typically, in this type of motor the field winding has a great many turns of fairly fine wire.

Fig. 9-4 *In a shunt motor, the field and armature windings are wired in parallel.*

When power is first applied to a shunt motor, the armature current is high, which translates to a fairly high torque. (Although, it will not be as high as the starting torque for a series motor.) As the motor comes up to speed, the counter-EMF increases until it is almost equal to the applied voltage. The speed stops increasing at this point and will remain relatively constant, almost independent of the mechanical load.

Applying a mechanical load tends to slow the motor down, of course. But it will also reduce the counter-EMF, which increases the torque, bringing the speed close to its no-load value.

To summarize, the shunt motor has a reasonable starting torque. Its biggest advantage is that its speed is nearly constant under varying load conditions. Obviously, the shunt motor is used primarily in applications where a constant speed is required.

Finally, there is the third type of dc motor, which is sort of a blend of the series motor and the shunt motor. This is the com-

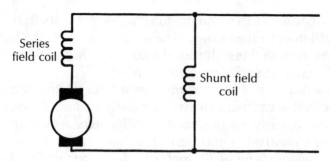

Fig. 9-5 *The compound motor has two field windings.*

pound motor. As shown in Fig. 9-5, this type of motor has two field windings. One is in series with the armature, while the other one is in parallel with it.

The compound motor behaves much like the series motor described earlier, with one important difference. The parallel field winding prevents the speed runaway problems that can be encountered with a series motor under no-load (or sometimes low-load) conditions. The parallel winding provides a nearly constant field strength. The speed will not try to increase without limit.

The speed, up to some definite maximum, is determined by the mechanical load. The compound motor is used for applications requiring a high torque over a varying load range.

Actually, there are two types of compound motors. The one just described is, by far, the most common. It is called the cumulative compound motor. The other type of compound motor is known as the differential compound motor. It behaves like a shunt motor and offers no particular advantage in the vast majority of applications, so it is rarely used.

ac motors

In some applications it may be desirable to run a motor off of an ac power supply, rather than a dc power supply. There are many different types of ac motors, and we won't go into detail here.

There are eight basic types of ac motors. They are

- split-phase
- capacitor-start
- two-value capacitor
- permanent-split capacitor

- shaded pole
- wound-rotor (repulsion)
- universal (or series)
- synchronous.

The main differences between the various motor types lie in the amount of starting torque they develop and their starting current requirements. The basic ac motor types are summarized in Table 9-1.

The universal motor

One additional type of motor should be mentioned here. The universal motor is rather unique in that it can be run on either ac or dc power. The operating characteristics of the universal motor are similar to those of the dc series motor. The starting torque is quite high. This type of motor is suitable for applications involving large loads and relatively short running times.

In most cases, a universal motor will run somewhat faster when it is operated on dc rather than ac power. This is because in dc operation, only the winding resistance is of significance, but in ac operation, both the resistance and the reactance of the windings affect the operation.

PROJECT 45:
Motor controller

Now that we have some basic familiarity with the various motor types, let's consider some of the ways in which they can be controlled. In this section we will only consider switching control. Speed control will be discussed later in this chapter.

To turn a motor on and off is easy enough. For manual control you simply use a switch to apply or remove power to the motor. In a remote control or automation system, it is just a matter of using the electronic equivalent of a switch. In most cases a transistor switch or a relay is employed. Figure 9-6 shows a typical relay-based motor control circuit. Certainly there should be nothing at all surprising in this.

In many control applications we may want to reverse the direction of the motor's rotation. For example, let's consider a door opening device. Let's say that the motor shaft rotates clock-

Table 9-1. Comparison of Various Types of ac Motors.

Type	Horsepower ranges	Load-starting ability	Starting current	Characteristics	Electrically reversible
Split phase	1/20 to 1/2	Easy starting loads. Develops 150% of full-load torque.	High; five to seven times full-load current.	Inexpensive, simple construction. Small for a given motor power. Nearly constant speed with a varying load.	Yes
Capacitor start	1/8 to 10	Hard starting loads. Develops 350% to 400% of full-load torque.	Medium; three to six times full-load current.	Simple construction, long service. Good general-purpose motor suitable for most jobs. Nearly constant speed with a varying load.	Yes
Two-value capacitor	2 to 20	Hard starting loads. Develops 350% to 450% of full-load torque.	Medium; three to five times full-load current.	Simple construction, long service, with minimum maintenance. Requires more space to accommodate larger capacitor. Low line current. Nearly constant speed with a varying load.	Yes
Permanent split capacitor	1/20 to 1	Easy starting loads. Develops 150% of full-load torque.	Low; two to four times full-load current.	Inexpensive, simple construction. Has no start winding switch. Speed can be reduced by lowering the voltage for fans and similar units.	Yes
Shaded pole	1/250 to 1/2	Easy starting loads.	Medium	Inexpensive, moderate efficiency, for light duty.	No
Wound rotor (repulsion)	1/6 to 10	Very hard starting loads. Develops 350% to 400% of full-load torque.	Low; two to four times full-load current.	Larger than equivalent size split-phase or capacitor motor. Running current varies only slightly with load.	No. Reversed by brush ring readjustment.
Universal or series	1/150 to 2	Hard starting loads. Develops 350% to 400% of full-load torque.	High	High speed, small size for a given horsepower. Usually directly connected to load. Speed changes with load variation.	Yes, some types.
Synchronous	Very small, fractional	N/A[1]	N/A	Constant speed.	N/A

[1]N/A = not applicable.

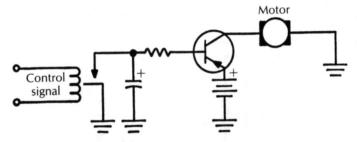

Fig. 9-6   *A relay can be used to control a motor.*

wise to open the door. In order for the same motor to close the door, it now has to rotate counterclockwise.

Reversing the direction of a dc motor is not difficult. It can be accomplished simply by reversing the direction of current flow in either the armature winding or the field winding, but not both. Reversing the direction of a dc motor with a manual DPDT switch can be accomplished with the circuitry illustrated in Fig. 9-7. A detail of the wiring of the switch is shown in Fig. 9-8. Figure 9-9 shows how the same thing can be accomplished with a pair of relays. One control signal causes the motor to run in one direction; a second control signal is used to operate the motor in the opposite direction.

While functional, this simplistic approach is far from ideal. Because two control signals are required, the control circuitry

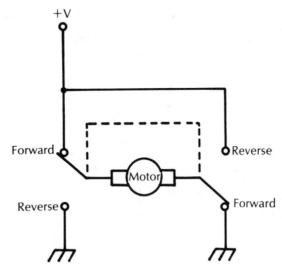

Fig. 9-7   *A DPDT switch can be used to manually reverse the direction of a motor's motion.*

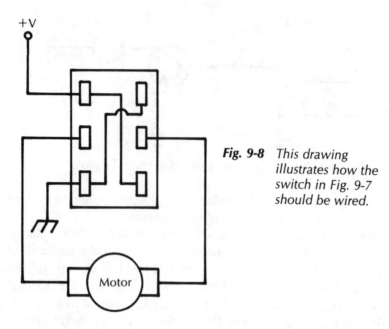

Fig. 9-8 *This drawing illustrates how the switch in Fig. 9-7 should be wired.*

can be complicated. An extra connecting wire might be required in a remote control application.

A potentially serious problem can occur if somehow both relays are activated at the same time. This is rather like a short circuit. The transistors and the motor can be damaged. Whenever multiple control signals are used, you must consider the results

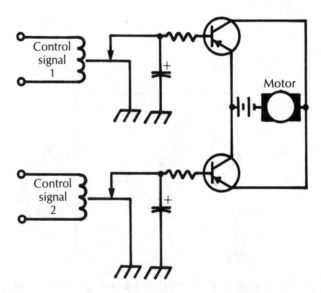

Fig. 9-9 *Two relays can also be used for bidirectional motor control.*

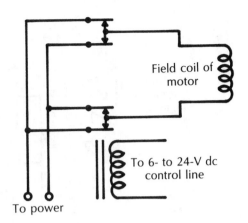

Fig. 9-10 *A DPDT relay can give bidirectional control of a motor with a single control signal.*

Field coil of motor

To 6- to 24-V dc control line

To power

of two or more appearing at once. Accidents can happen. For example, consider a remote control application with two signal wires and a common wire in the connecting cable. If the two signal wires are shorted together somewhere along their length, a control signal on one will appear on the other.

Back in Fig. 9-7 we manually controlled the direction of the motor with a DPDT switch. Why can't we use a DPDT relay to accomplish the same ends? The answer is, there is no reason at all why not. Figure 9-10 illustrates how this might be done.

A rather different approach to controlling motor direction is illustrated in Fig. 9-11. Here a digital logic signal is used to control the motor. A logic 0 (low) drives the motor in the forward direction. A logic 1 (high) reverses the motor's rotation. Switching is accomplished via four VMOS-type FETs (field-effect transistors). The parts list for this project is given in Table 9-2.

In certain control applications, we don't want the motor shaft to rotate freely. Some limitation of its motion may be required. As an example, let's say we have a control system that mechanically turns a potentiometer, such as a volume control. A potentiometer turns just so far before it reaches the limit of its motion. If the motor keeps turning and tries to force the potentiometer past its normal limit, damage will result.

One potential solution to this type of problem is to use a loose mechanical coupling between the motor and the potentiometer. When the potentiometer reaches one of its extreme positions, it will resist further movement in that direction. The loose mechanical linkage will slip, allowing the motor to continue turning without forcing the potentiometer past its limits. Unfortunately, once the coupling has slipped a few times, it may be-

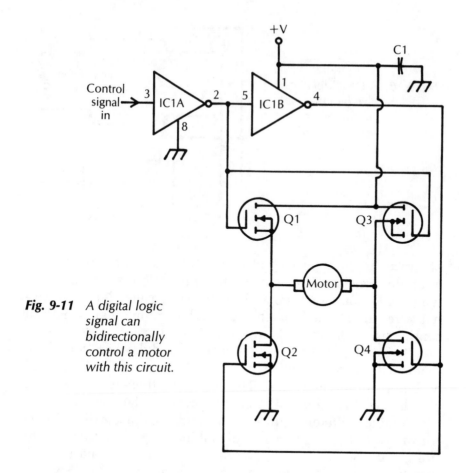

Fig. 9-11 *A digital logic signal can bidirectionally control a motor with this circuit.*

Table 9-2 Parts List for the Digital Motor Control of Fig. 9-11.

IC1	CD4049 hex inverter IC
Q1 – Q4	VMOS FET (VN (VN67 or similar)
C1	0.1-μF capacitor

come too loose and the linkage may become erratic. Eventually the motor won't be able to turn the potentiometer at all because the coupling will be too loose to catch.

A better approach is shown in Fig. 9-12. A pair of NC (normally closed) snap action switches are used to mark the mechanical limits of motion. The associated circuitry is illustrated in Fig. 9-13.

When the limit is reached in one direction, switch S1 opens. This allows a gate current to reach the SCR, turning it on. The

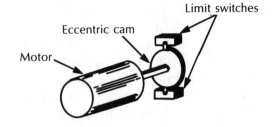

Fig. 9-12   *Snap-action switches can be used to limit mechanical motion.*

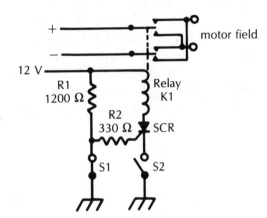

Fig. 9-13   *This is the circuitry for the system illustrated in Fig. 9-12.*

relay is activated, reversing its switching contacts. The motor now starts to rotate in the opposite direction.

As soon as the motor starts moving away from the limit position, the limit switch is allowed to close, removing the gate signal from the SCR. This has no effect on the SCR, which continues to conduct until the anode/cathode is interrupted.

At the opposite limit position, switch S2 is opened. This interrupts the anode/cathode current flow through the SCR. It now shuts off, deactivating the relay. The switching contacts revert to their original position. The motor now starts turning in the original direction. While this system has its limitations, it will prove useful in a number of control applications.

PROJECT 46:
Speed control

In many control applications it may be important to control the speed as well as (or instead of) the direction of the motor's rotation. Controlling motor speed is a little more difficult than controlling direction, but it is still far from impossible. There are several possible approaches.

To a large extent the speed of a motor is determined by the applied voltage. This suggests that you can control the speed with a simple voltage divider circuit built around a potentiometer or rheostat. In some cases this will work. In most instances, however, it will result in very unreliable operation.

One problem is that most motors draw a fairly large amount of current. The power drawn through the potentiometer can literally go up in smoke. High-power potentiometers and rheostats can be used, but they tend to be expensive and bulky. Besides, operation can still be unreliable.

The culprit is inertia. It takes more energy to start a motor moving than it takes to keep it moving. At low speed settings with a potentiometer, there might not be enough power to over-

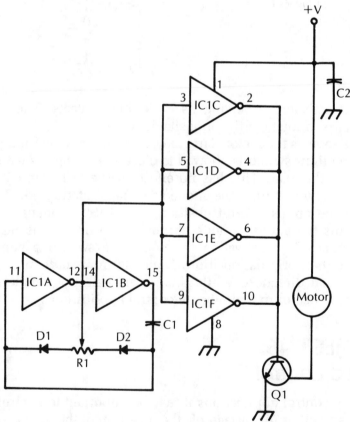

Fig. 9-14 *A variable frequency square-wave generator can be used to control the speed of a motor.*

come the initial inertia. It can be done, with a little more effort. The potentiometer can be set for a relatively high speed to start the motor, then backed off to the desired low speed. This is inefficient at best, and can be difficult to accomplish conveniently in a control system.

Instead of using voltage control to set the speed, a better approach is to use duty-cycle control. A driving oscillator is used instead of dc voltage to power the motor. The oscillator puts out a rectangular wave that switches repeatedly between V+ and ground. In other words, the V+ signal to the motor is switched on and off at a rapid rate (many times a second). The ratio between the on time and the off time is called the duty cycle. The amount of on time determines the speed of the motor.

By adjusting the frequency of the oscillator, we can control the speed. A simple circuit for accomplishing this is illustrated in Fig. 9-14. The parts list is given in Table 9-3.

Table 9-3   Parts List for the Motor Speed Control of Fig. 9-14.

IC1	CD4049 hex inverter IC	
Q1	NPN transistor (2N3055 or similar)	
D1, D2	Diode (1N914 or similar)	
R1	1-MΩ potentiometer	
C1	0.022-μF capacitor	
C2	0.1-μF capacitor	

By using duty-cycle control, we make inertia work for us instead of against us. The motor is turned on at full speed, overcoming its starting inertia. During the off portion of the oscillator signal, the motor is turned off. But inertia works both ways. It will take some finite time before the motor comes to a complete stop. Before it gets a chance to slow down significantly, the control signal goes high again, turning on the motor full speed. The end result is a fairly smooth, constant speed that is proportional to the oscillator frequency.

In the circuit of Fig. 9-14, the oscillator frequency is set via potentiometer R1. Almost any oscillator circuit can be used in this type of application.

The 2N3055 transistor called for in the parts list should be sufficient for low-power motors. A larger, more powerful transistor might be required to drive a larger motor. The transistor must be able to handle the current drawn by the motor. Check the manufacturer's spec sheets for both the motor and the transistor.

Electronic switching

YOU HAVE PROBABLY NOTICED THAT CERTAIN BASIC PRINCIPLES continuously crop up in control applications. Practically all remote control and automation devices involve some form of electronic switching. This chapter will explore several important aspects of electronic switching.

Relays

Probably the simplest form of electronic switching is the relay. We have used these devices in many of the projects already presented in this book. Just what is a relay anyway? Basically, a relay consists of two main parts: a coil and a magnetically controlled switch. When current flows through the coil, a magnetic field builds up around it. This magnetic field activates the switch. The magnetic field attracts an armature in the switch. The movement of the armature makes or breaks one or more switch contacts. When current stops flowing through the coil, the magnetic field collapses. A spring pulls the armature back to its original position and the switch contacts revert to their original states.

Like manual switches, relay switch contacts come in a variety of configurations, as illustrated in Fig. 10-1. The simplest version has just two contacts. This is called an SPST (single pole, single throw) switch. The contacts can be normally open (NO) or normally closed (NC). The ''normal'' state is the condition of the switch contacts when the relay is not activated (no current flowing through the coil).

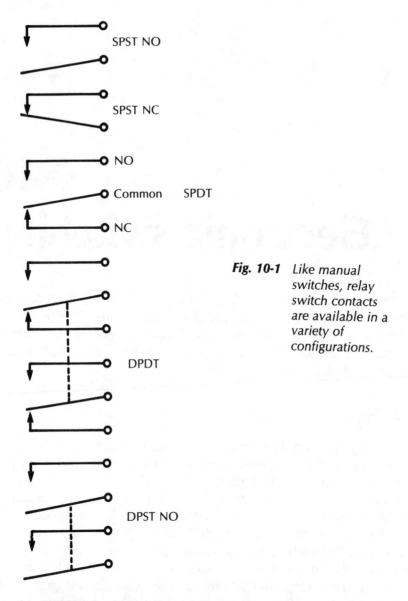

SPST NO

SPST NC

NO

Common   SPDT

NC

DPDT

DPST NO

Fig. 10-1 *Like manual switches, relay switch contacts are available in a variety of configurations.*

An SPDT (single pole, double throw) switch has three contacts. The middle contact (armature) is the common. One of the other contacts is normally open and the other is normally closed. When the relay is activated, one contact is made and another is broken.

The next more complex switch contact arrangement is the DPDT (double pole, double throw). The DPDT switch is essentially two independent SPDT switches that are operated in unison.

DPST (double pole, single throw) switches are also possible, but are rarely used. If you need a DPST format, use a DPDT and leave the NC contacts open. Multiple (more than two) poles or throws are also available for complex switching applications.

The coil requirements are of considerable importance. Some are designed to work on ac, others on dc. The activating voltage ranges from 1 to 250 V. Commonly available relays generally tend to use just a few standard values, including

- 6 V
- 12 V
- 24 V
- 48 V
- 117 V
- 240 V.

Low-voltage relays (under 100 V) are usually dc types. The higher voltage ratings (117 V and 240 V) are usually ac models.

The applied voltage should be close to the rated value, although it doesn't need to be exact. Generally, if you are within 20% (plus or minus) of the rated value, you won't have any problems. Too high a voltage can burn out the coil or damage the armature. Too low a voltage probably won't do any damage, but it could result in unreliable switching.

Whenever you have a switch in series with an inductance (coil), there is a potential for trouble. When the switch is opened (voltage removed), the magnetic field around the coil collapses. A voltage proportional to the rate of change of current is self-inducted in the coil. This voltage can be quite high because the current drops to zero very rapidly. This high voltage can eventually damage the relay. To protect against such damage, it is a good idea to place a diode across the relay coil, as shown in Fig. 10-2.

The diode's PIV (peak inverse voltage) rating should be higher than the power supply voltage. The current handling capability should be considerably greater than the load operating

Fig. 10-2 *A diode should always be placed across the relay coil to protect against high-voltage self-inductance spikes.*

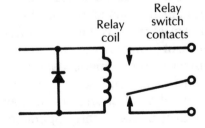

Relay coil

Relay switch contacts

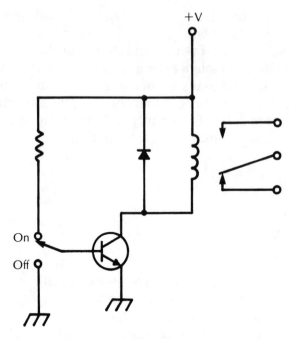

Fig. 10-3 *A high-current relay can be driven from a low-current source by using a simple transistor amplifier.*

current. A current-handling capability of 20 to 30 times the load operating current is appropriate.

In most of the projects in this book, I have used a 1N4002 diode, which is readily available. Its PIV and current ratings are considerably greater than required for any of the circuits.

In some applications, you will need to drive a high-current relay from a low-current source. You can cascade two relays, but generally it is more economical to use a simple transistor amplifier, like the one shown in Fig. 10-3.

Transistor switches

One of the chief advantages of using electronic circuitry is that there are no mechanical parts. Mechanical parts wear out and perform erratically far more often than electronic components. They can jam, corrode, get dirty, break, bend, or develop any of a number of problems, especially when there are relatively small and delicate parts.

A relay, of course, is primarily a mechanical device, even though it is electrically operated. They are reasonably reliable, but are often the first part in a circuit to develop problems. In addition, while fast, a relay does take a finite amount of time to change its switch connections. When several relays are involved in a system, the cumulative time delay can add up to a significant total.

Fortunately, semiconductor devices can be used in electronic switching applications. They are small, reliable, inexpensive, and very fast. They also consume a relatively low amount of power, as compared to relays.

Bipolar transistors make excellent electronic switches. To get a transistor to function as an electronic switch, it must be correctly biased. There are three basic biasing schemes, or modes, for transistor switching circuits. They are as follows:

- the saturated mode
- the current mode
- the avalanche mode.

We will discuss each of these modes briefly.

Saturated mode

In the saturated mode, the transistor is turned on by biasing it into saturation. The collector current is limited only by the resistances in the collector and emitter circuits. The voltage across the transistor (known as the saturation voltage) is at a minimum. The exact level of the saturation voltage is defined by the collector current and the load resistance. The off condition is achieved by biasing the transistor so that it is cut off; that is, so no collector current flows.

A simple saturation-mode switching circuit is shown in Fig. 10-4. Battery (voltage source) V_{BB} biases the transistor into cut-off when no input signal is present. The base is made negative with respect to the emitter. The transistor is off. If a sufficiently positive voltage is applied to the input, it will overcome this negative bias voltage, switching the transistor on, and allowing collector current to flow.

Voltage will be dropped across the load resistor (R_L) only when there is some collector current. Since there is a collector current only when the transistor is turned on (positive voltage at the input), there will be a voltage drop across R_L only when the transistor switch is in its on condition. Otherwise, the output voltage will be zero.

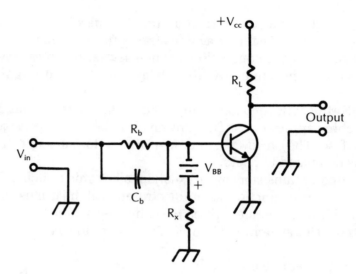

Fig. 10-4 *This is a basic saturation-mode switching circuit.*

Ideally, the transistor should switch on and off instantly, with no transition time between states. This simply isn't possible with practical real-world devices. Some finite time is required for the transistor to reverse its output state. The switching time is usually measured in microseconds, which is close enough to an instant response for the vast majority of practical applications. A transistor's switching time is considerably faster than the switching time of a relay.

Let's assume we are working with the circuit shown in Fig. 10-4. Initially, it is in its off state. There is no collector current, so there is no voltage across R_L.

Now, at some specific time (t_0), a positive pulse is applied to the input. Base current starts to flow right away, but there is a brief period of time before the emitter/base voltage climbs from its initial negative value through zero to a positive voltage. Collector current cannot even begin to flow until the emitter/base voltage is at least slightly positive. Then the collector current will require some finite time to reach its maximum level from the starting point of zero.

The turn-on time is considered to be the time from when the positive voltage is first applied to the base (t_0) and the instant when the collector current reaches 90% of its maximum value (I_1). This is called the delay time. It is almost always very small. Similarly, the transistor cannot go instantly from saturation to cutoff when the input voltage is removed.

When the positive input voltage is removed, V_{BB} brings the voltage on the base back to a negative level. The base current momentarily goes negative until the emitter/base voltage goes negative, and the emitter/base junction ceases to conduct. The emitter/base voltage and the collector current remain positive for a brief period after the base is brought down to a negative voltage.

The time it takes for the collector to drop to 10% of its maximum level after the positive input signal is called the storage time, and it is determined by the internal capacitances formed within the transistor while it is in saturation. These capacitances are charged when the transistor is turned on and discharge relatively slowly when the transistor is turned off. Because these internal capacitances are very small, we are generally talking about storage times in the microsecond or millisecond range.

Bridging capacitor C_b across the base resistor R_b increases the switching speed. This capacitor is not always used because it is not essential for operation of the circuit.

To find the component values in this circuit you must know the saturation current $I_{c(sat)}$. This value can be easily found via Ohm's law:

$$I_{c(sat)} = \frac{V_{CC}}{R_L}$$

Resistance R_L is selected to match the output load being driven by the circuit.

Let's assume the following values are in the circuit:

- V_{CC} = 9 V
- V_{BB} = 1.5 V
- R_L = 10 kΩ (10,000 Ω).

In this case, the saturation current works out to

$$I_{c(sat)} = \frac{9}{10,000} = 0.0009\ A = 0.9\ mA$$

The value of resistor R_x is found using this form of Ohm's law:

$$R_x = \frac{V_{BB}}{I_{CBO}}$$

where

V_{BB} = negative base voltage and

I_{CBO} = collector-to-base leakage current when the transistor is operating at its maximum temperature.

This value can be obtained from the manufacturer's specification sheet for the transistor used. If we assume I_{CBO} is 2 μA (0.000002 A), R_x should have a value of

$$R_x = \frac{1.5}{0.000002} = 750,000 \ \Omega = 75 \ \text{k}\Omega$$

Next, we need to look at a characteristic curve graph for the transistor (included in the specification sheet). A typical characteristic curve graph is shown in Fig. 10-5.

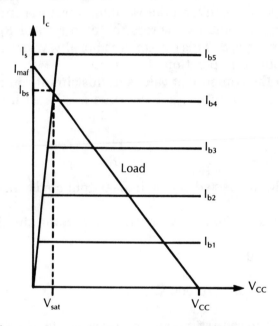

Fig. 10-5 *This is a characteristic curve graph illustrating a transistor's operating characteristics under various conditions.*

Plot the load line for the collector circuit (the solid line marked "Load" in the graph). An estimated value for the base current (I_{bs}) is indicated by the point where the load line crosses the transistor's saturation resistance curve. In our example, this is just a little greater than I_{b4}.

We can now find the value of the base resistor (R_b):

$$R_b = \frac{V_{IN(max)}}{I_{bs} - I_{CBO}}$$

where

$V_{IN(max)}$ = maximum input voltage.

Let's say that I_{BS} works out to 0.225 mA (0.000225 A), and the maximum input voltage is 2.50 V. In this case R_b should have a value approximately equal to

$$R_b = \frac{2.50}{0.000225 - 0.000002}$$
$$= \frac{2.5}{0.000223}$$
$$= 11211 \ \Omega$$
$$\approx 12,000 \ \Omega = 12 \ k\Omega$$

If capacitor C_b is used, its value can be derived mathematically, but it is usually easier just to breadboard the circuit and experiment with different capacitance values until the best switching speed is obtained.

Current mode

Higher switching speeds can be obtained if the transistor is not put into saturation when it is turned on. In the current mode, the transistor is biased so that it operates close to, but not quite in, saturation. The collector-emitter voltage is therefore somewhat greater than the saturation voltage of the device. As in the saturated mode, the transistor switch is turned off by biasing it so that the semiconductor is in the cutoff (nonconducting) state.

A typical current-mode switching circuit is illustrated in Fig. 10-6. Notice that it is very similar to the saturated-mode circuit of Fig. 10-4.

The battery (V_{BB}) and resistor (R_x) combination in the base circuit holds the transistor cutoff when there is no input signal. This works essentially the same as the saturated-mode circuit described earlier.

The emitter voltage source (V_{EE}) keeps diode D1 turned on (conducting) at all times. Assuming a silicon diode is used, there will be approximately 0.7 V across this component. If the base of the transistor is at ground potential (0 V), only the 0.7 V across

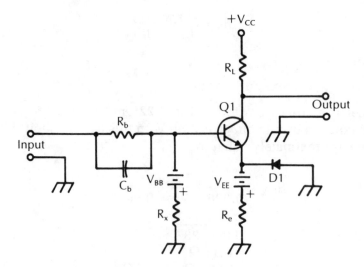

Fig. 10-6 *A current-mode switching circuit is similar to, but faster than, a saturated-mode switching circuit.*

the diode will be between the emitter and the base. This will keep the transistor conducting with 0 V at the base.

The base bias voltage (V_{BB}) has to be canceled out by an input signal with the opposite polarity. When V_{in} is equal to or greater than V_{BB}, the transistor is switched on. When V_{in} is less than V_{BB}, the transistor is cut off.

If V_{EE} was not included in the circuit, we would have practically the same situation as with the saturated-mode circuit of Fig. 10-4. When the transistor is turned on by a positive pulse at the input, the collector is equal to the beta of the transistor multiplied by the base current. If the base current is large enough, the transistor is driven into saturation.

The three added components in the emitter circuit (V_{EE}, R_e, and D1) limit the input current and, therefore, the collector current. The maximum current through the transistor is less than the saturation value. The maximum current flow through this circuit's transistor is determined by the component values in the emitter circuit.

The diode's polarity prevents any current from the emitter from flowing through it, but the 0.7 V across the diode does limit the current flow through R_e, which is the maximum current that can flow through either the emitter or the collector in this circuit. The transistor will not be driven into saturation because the saturation current value is greater than the current through R_e. In practical circuits, the current through the load resistor (R_L) should be limited to a maximum level equal to the current through the emitter resistor (R_e).

In our saturated-mode example in the last section, we found that the saturated collector current (the transistor's on current) was 0.9 mA. For a current-mode circuit, we want the emitter limiting current ($I_{el} = V_{EE}/R_e$) to be less than this value. Let's assume we want a value of 0.7 mA (0.0007 A). If V_{EE} is 1.5 V, the emitter resistor should have a value of about

$$R_e = \frac{V_{EE}}{I_{el}} = \frac{1.5}{0.0007} = 2,143 \ \Omega$$

If we use a standard 2.2-kΩ (2,200-Ω) resistor, the maximum collector current/emitter current will be approximately

$$I_{max} = \frac{1.5}{2,200} = 0.00068 = 0.68 \ \text{mA}$$

The transistor will not be put into saturation when it is turned on.

Because the transistor does not have to work quite so hard (conduct as much current) in its on state, it can switch between states considerably faster than in a comparable saturated-mode switching circuit.

Avalanche mode

The third switching mode is also the fastest. This is the avalanche mode. The on and off states of the transistor switch are kept within the breakdown portion of the transistor's operating curve. Faster switching can be obtained with special devices such as hot-carrier, pin, snap-off, or tunnel diodes.

Saturated-mode and current-mode switching circuits require a specific base voltage to be maintained to hold the transistor off (or on). In an avalanche-mode circuit, however, a brief pulse is all that is needed to hold the transistor in either the on or off state. A continuous input voltage signal is not required.

Other semiconductor switching devices

In the last section we considered switching circuits built around bipolar transistors. Other semiconductor devices, such as FETs,

UJTs, PUTs, SCRs, and triacs can also be employed in electronic switching applications.

FET switches

The FET (field-effect transistor) is a semiconductor device with operating characteristics similar to those of a vacuum tube. FETs have better on-off current ratios than bipolar transistors. However, FET switches operate somewhat slower than circuits built around bipolar devices. This is because of the large internal capacitances of a FET. A switching circuit built around an FET is illustrated in Fig. 10-7.

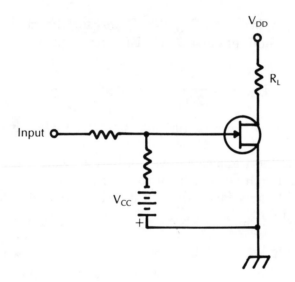

Fig. 10-7 *FETs can also be used for switching applications.*

UJT switches

Another variation on the transistor is the UJT (unijunction transistor). It too can be used in switching applications. An ordinary (bipolar) transistor has two PN junctions. A unijunction transistor, on the other hand, has just one. Its three leads include an emitter and two base connections. A UJT is quite simple, as shown in Fig. 10-8.

An alternate form of the UJT is the PUT (programmable unijunction transistor). It is very similar to the UJT, except more of its operating characteristics are determined by the external circuitry, offering greater flexibility to the circuit designer.

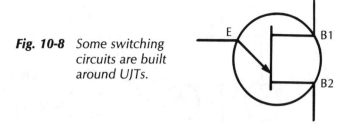

Fig. 10-8 *Some switching circuits are built around UJTs.*

SCR switches

Many switching circuits use the SCR (silicon controlled rectifier). This device is essentially a gated diode; that is, a diode that can be electrically turned on and off. Like an ordinary diode, the SCR has an anode and a cathode. It also has a third lead called the gate.

The SCR will not conduct until a triggering pulse is applied to the gate. Current starts flowing from the anode to the cathode. Current continues to flow, even after the gate signal is removed. Once the SCR has been turned on, the only way it can be turned off is to break the anode/cathode current path or reverse the direction of the current flow. The SCR, like the diode, conducts in only one direction. It will block current of the opposite polarity.

In some circuits this monopolarity can be very useful. For example, in many light dimmers, the SCR is used to cut off part of each ac cycle, as shown in Fig. 10-9. When the signal exceeds a specific switching level, the gate is activated and the SCR starts to conduct. When the ac cycle goes into its negative portion, the SCR is cut off. This whole process limits how much of the ac power reaches the load.

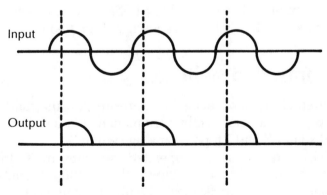

Fig. 10-9 *An SCR can be used to delete a portion of an ac waveform.*

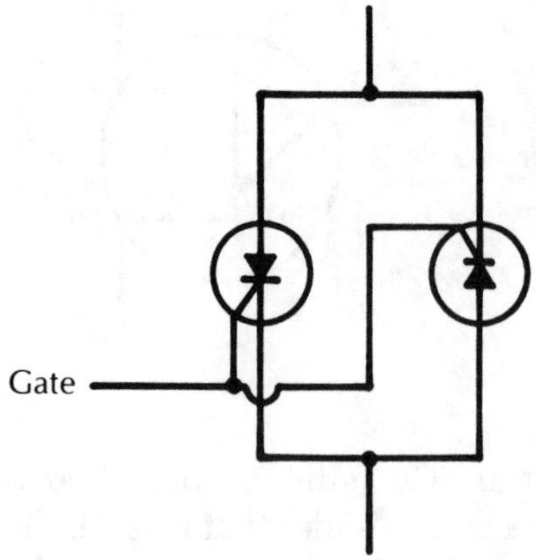

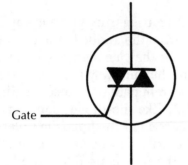

Fig. 10-10   *Two SCRs can be connected like this for bipolar operation.*

Fig. 10-11   *A triac is essentially a pair of back-to-back SCRs in a single housing.*

In other applications, the monopolarity of the SCR can be problematic. If we need to conduct current in both directions, we can place two SCRs in reverse polarity, as illustrated in Fig. 10-10. An even more convenient solution is to use a triac. This device is essentially a pair of back-to-back SCRs in a single housing. The symbol for a triac is shown in Fig. 10-11.

dc Controlled switches

An extremely handy device has appeared on the market in the last few years. It is known by various names, including "dc controlled switch" and "bipolar analog switch."

These are electrically operated switches in IC form. The most common device of this type is the CD4066. It contains four independent switching stages. The pin diagram is shown in Fig. 10-12.

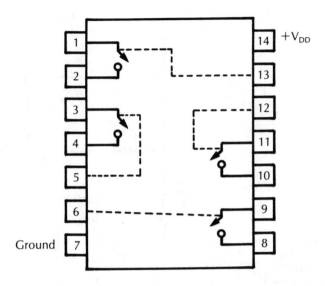

Fig. 10-12 *The CD4066 IC contains four independent dc-operated switches.*

While the CD4066 is a CMOS-type digital IC, it can be used with either digital or analog circuitry. A digital control signal can be used to operate a switch carrying an analog signal in either direction. The switches are not polarity sensitive.

PROJECTS 47 and 48:
Touch switches

One of the most unexpected switching devices is the touch of a finger, or foot, or whatever. Touch control circuits can be used to control almost anything with just the lightest touch. A touch switch can be easily operated when your hands are full, or it can be used in many hidden sensor applications.

A simple touch switch circuit is illustrated in Fig. 10-13. The parts list is given in Table 10-1. There is a lot of flexibility in this circuit. Nothing is terribly critical. Almost any N-channel FET and NPN transistor can be used. The transistor must be able to comfortably handle the current drawn by the load.

The touch plate is an exposed metallic contact. Obviously it must be 100% isolated from any ac power source for safety. Only low-power dc voltages flow through the touch switch circuit itself. Potentiometer R4 is used to adjust the shutoff time of the circuit. Experiment with settings.

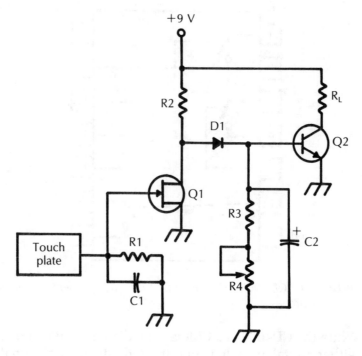

Fig. 10-13   *Almost anything can be controlled by a simple touch with this circuit.*

Table 10-1. Parts List for the Touch Switch of Fig. 10-13.

Q1	N-channel FET (see text)
Q2	NPN transistor (see text)
D1	1N914 diode
C1	100-pF capacitor
C2	15-μF, 25-V electrolytic capacitor
R1	10-MΩ resistor
R2	2.7-kΩ resistor
R3	22-kΩ resistor
R4	50-kΩ potentiometer
R_L	Load resistance

I strongly recommend that you breadboard this circuit before constructing a permanent version. While the component values are not especially critical, sometimes this type of circuit can be a little fussy, and it might require some minor changes or fine tuning before it operates reliably.

A second touch switch circuit is shown in Fig. 10-14. The parts list for this project is given in Table 10-2.

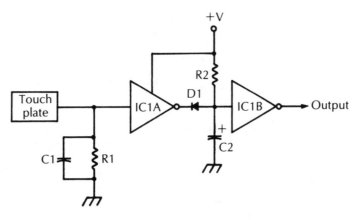

Fig. 10-14 *This is an alternative touch switch circuit.*

Table 10-2   Parts List for the Touch Switch of Fig. 10-14.

IC1	CD4049 hex inverter	
D1	1N914 diode	
C1	100-pF capacitor	
C2	1-μF capacitor	
R1	10-MΩ resistor	
R2	100-kΩ resistor	

How do these circuits work? Sixty-hertz power signals are almost always all around us. Low-level 60-hertz signals are picked up by the body and can be transmitted through a fingertip (or other body part) to a small touchplate. In some cases, a simple bare end of wire will do, but usually a larger plate is more convenient, and possibly more reliable. I often use a small piece of copper clad circuit board (the type used for homemade PC boards), but almost anything conductive will do the trick.

With either of these projects, or any touch switch circuit, you should use battery power only. Do not use an ac-to-dc power supply. There is always a possibility of a short forming that could permit live ac at dangerous current levels to reach the touch plates. Please, do not take foolish chances.

PROJECT 49:
Timed touch switch

This project is yet another touch switch, but with a difference. When this touch switch is activated, by touching it briefly, it stays

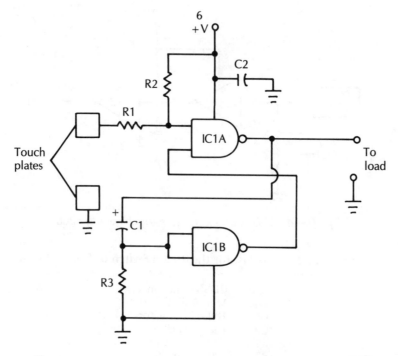

Fig. 10-15 *This touch switch circuit features a built-in timer.*

on for a fixed period of time, then automatically switches itself off, until it is touched again. The length of the activating touch is irrelevant. The schematic diagram for this timed touch switch is shown in Fig. 10-15, and the parts list is given in Table 10-3.

Basically, what we have here is a touch switch triggering a simple monostable multivibrator made up of a pair of NAND gates (IC1A and IC1B). The remaining two sections of the CD4011 quad NAND gate can be used in other circuitry as part of a larger system. If they are not used, all unused inputs and outputs should be grounded. A gloating input on a CMOS chip can cause instability problems in the other (used) sections.

Normally, the output of this circuit is low. The circuit is activated by shorting a pair of touch plates with your finger, as shown

Table 10-3 Parts List for the Timed Touch Switch of Fig. 10-15.

IC1	CD4011 quad NAND gate
C1	5-μF, 25-V electrolytic capacitor
C2	0.01-μF capacitor
R1	10-MΩ, 0.25-W resistor
R2, R3	120-kΩ, 0.25-W resistor

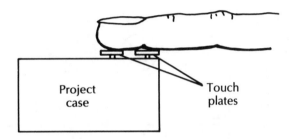

Fig. 10-16 *The timed touch switch circuit of Fig. 10-15 is activated by bridging a pair of touch plates with a fingertip.*

in Fig. 10-16. These touch plates should be positioned very close to each other to permit easy and convenient triggering.

When the touch plates are shorted, even for a brief instant, the circuit's output goes high for a specific period of time, then it goes low again, even if the touch plates are still being shorted. In other words, the output pulse will always be the same length. The length of the activating touch makes no difference to the operation of the circuit.

For the component values suggested in the parts list, the timing period is approximately 1 second. You are encouraged to experiment with other component values in this circuit. The main timing components are resistor R3 and capacitor C1. It is a good idea to keep resistors R2 and R3 at more or less equal values. Resistor R1 should have a significantly higher value than resistors R2 or R3. At a bare minimum, R1's value should be at least 10 to 20 times the resistance of R3. Capacitor C2 protects the CMOS gates from possible transients in the power supply lines.

As with any touch switch circuit, this project should be powered from batteries only. Do not use an ac-to-dc power supply. There is always a possibility of a short forming that could permit live ac at dangerous current levels to reach the touch plates. Please, do not take foolish chances.

PROJECT 50:
Overvoltage protector/electronic crowbar

The circuit shown in Fig. 10-17 provides fast and efficient protection against dangerous short circuits or other overvoltage conditions. If the supply voltage exceeds a specific trigger point, power to the load circuit is shut down within a fraction of a sec-

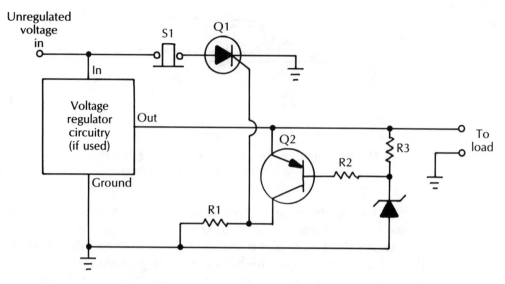

Fig. 10-17 *This is an overvoltage protector/electronic crowbar circuit.*

**Table 10-4   Parts List for the
Overvoltage Protector/Electronic Crowbar of Fig. 10-17.**

Q1	SCR (to suit load)	
Q2	PNP transistor (2N3702 or similar)	
D1	Zener diode (select for voltage trigger point) (see text)	
S1	NC SPST push-button switch	
R1	1-kΩ, 0.5-W resistor	
R2	4.7-kΩ, 0.5-W resistor	
R3	100-Ω, 0.5-W resistor	

ond. This does not absolutely guarantee that no damage will be done to the load or any of its components, but it will minimize the risk significantly. The parts list for this project is given in Table 10-4.

The SCR should be selected to carry enough current to supply the intended load. It is a good idea to overrate the SCR by at least 25%. For example, if your intended load is designed to draw up to 1 A, the SCR should be rated for 1.25 A or more to provide a little extra headroom.

The trigger point for this electronic crowbar is determined by zener diode D1. For example, if a 6.2-V zener diode is used, the load's power supply will be shut down if the supply voltage ever goes above 6.2 V.

Almost any PNP transistor can be used for Q2. This transistor does not have to carry the full load current because it is only part of the triggering network for the SCR's gate.

PROJECT 51:
Remote resistor selector

Most simple remote control systems use simple on/off signals. In some practical applications, however, there might be a need to remotely control some other electrical variable. A simple, but very useful, remote resistor selector circuit is shown in Fig. 10-18. This is not really a complete project in itself. It is intended to be used as part of a larger system. A typical parts list for this

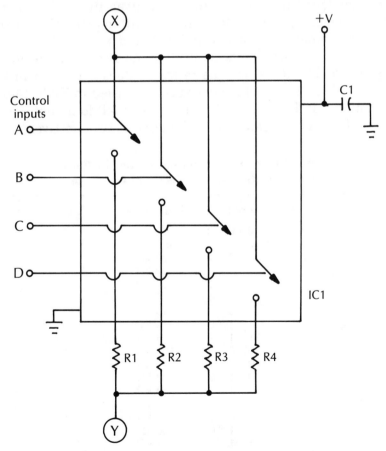

Fig. 10-18   *This is a remote resistor selector circuit.*

Table 10-5 Parts List for the Remote Resistor Selector of Fig. 10-18.

IC1	CD4066 quad bilateral switch
C1	0.01-μF capacitor
R1	100-kΩ resistor (see text)
R2	68-kΩ resistor (see text)
R3	39-kΩ resistor (see text)
R4	27-kΩ resistor (see text)

project is given in Table 10-5. The actual resistor values used will depend on the specific requirements of your particular intended application, of course.

Four binary (high/low) signals control the four semiconductor switches in a CD4066 quad bilateral switch chip (IC1). These switches determine which of the four resistors (R1 through R4) will be switched into the circuit at any given time. This circuit can then be incorporated into a larger circuit in place of any fixed resistor by connecting it across points X and Y.

With four binary inputs controlling the selector, there are 16 possible combinations. For the resistor values suggested in the parts list, these 16 possible resistances are listed in Table 10-6.

Notice that if all four switches are open (all four control signals are low), no resistors are in the circuit. We effectively have

Table 10-6 Available Resistance Combinations for the Remote Resistor Selector of Fig. 10-18.

Inputs				Resistance
A	B	C	D	
0	0	0	0	(infinity)
0	0	0	1	27 kΩ
0	0	1	0	39 kΩ
0	0	1	1	16 kΩ
0	1	0	0	68 kΩ
0	1	0	1	19 kΩ
0	1	1	0	24 kΩ
0	1	1	1	13 kΩ
1	0	0	0	100 kΩ
1	0	0	1	21 kΩ
1	0	1	0	28 kΩ
1	0	1	1	14 kΩ
1	1	0	0	40 kΩ
1	1	0	1	16 kΩ
1	1	1	0	20 kΩ
1	1	1	1	11 kΩ

an open circuit under this condition. This gives us a nominal resistance of infinity. Actually, the resistance in this case will be very high, but not truly infinite. It will be limited by the internal off resistance of the IC itself.

In some applications, we might want to limit the total maximum resistance to a specific level. Just connect another large-value resistor in parallel across points X and Y. This resistor's value will be the maximum resistance seen by the load circuit. Of course, this extra paralleled resistor will affect the other resistance values for each switching combination, but if this fifth resistor's value is very large in comparison with R1 through R4, the effect can be made fairly negligible.

You should notice that the resistances do not increase linearly as the binary control value is increased. Instead, the resistance goes up and down. That is because when more than one of IC1's switches is on, we have two (or more) resistors in parallel. The combined effective value of resistances in parallel is always less than any of the individual component resistances in this combination.

In a remote control system, five lines are required between the remote control station and this circuit—a common ground

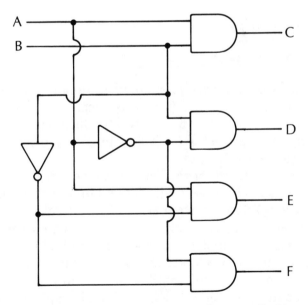

Fig. 10-19   *The circuit shown in Fig. 10-18 can be controlled by two digital signal lines.*

**Table 10-7   Action of the Gating Circuit
for the two-to 4-line Demultiplexer of Fig. 10-19.**

Inputs		Outputs			
A	B	C	D	E	F
0	0	0	0	0	1
0	1	0	1	0	0
1	0	0	0	1	0
1	1	1	0	0	0

and the four binary control lines. In some cases, this can be reduced to two control lines (and the common ground) by multiplexing the binary signals. A simple two-line to four-line demultiplexing network is illustrated in Fig. 10-19. The action of this gating circuit is summarized in Table 10-7.

The disadvantage of using this demultiplexer is that only one of the four switches can be activated at any time. Parallel combinations are not possible.

Of course this circuit can easily be expanded. Adding a second CD4066 quad bilateral switch IC permits control of up to eight resistors. Of course, four more control lines will be needed.

PROJECT 52:
Remote capacitor selector

The previous project can be adapted to remotely control other electrical variables besides just resistance. A remote capacitor selector circuit is shown in Fig. 10-20. Once again, this is not really a complete project in itself. It is intended to be used as part of a larger system.

A typical parts list for this circuit is given in Table 10-8. The actual capacitor values used will depend on the specific requirements of your particular intended application, of course.

Four binary (high/low) signals control the four semiconductor switches in a CD4066 quad bilateral switch chip (IC1). These switches determine which of the four capacitors (C1 through C4) will be switched into the circuit at any given time. This circuit can then be incorporated into a larger circuit in place of any fixed capacitor by connecting it across points X and Y.

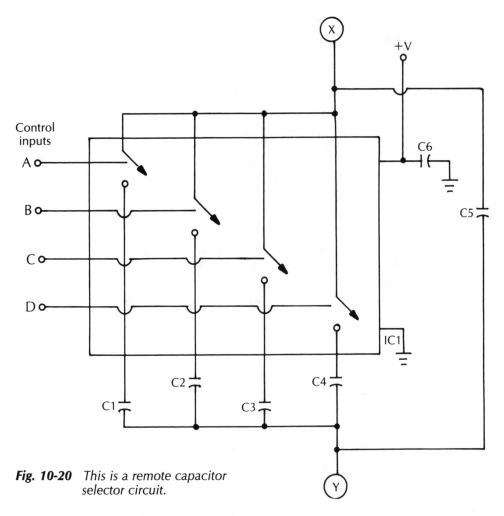

Fig. 10-20 *This is a remote capacitor
selector circuit.*

A fifth capacitor (C5) is permanently in parallel with the
entire switching circuit to ensure against an open circuit condition. The value of this capacitor is small in comparison with the
four main switch capacitors (C1 through C4).

With four binary inputs controlling the selector, there are 16
possible combinations. For the capacitor values suggested in the
parts list, these 16 possible resistances are listed in Table 10-9.

Notice that, unlike the remote resistor selector discussed in
the last section, here the capacitance values increase steadily as
the four-digit binary control value is increased. This is because
capacitances in parallel are added together. Additional information on this type of circuit can be found in the text for project 51.

**Table 10-8   Parts List for the
Remote Capacitor Selector of Fig. 10-20.**

IC1	CD4066 quad bilateral switch
C1	0.1-μF capacitor (see text)
C2	0.05-μF capacitor (see text)
C3	0.0.2-μF capacitor (see text)
C4	0.01-μF capacitor (see text)
C5	0.005-μF capacitor (see text)
C6	0.01-μF capacitor

**Table 10-9   Available Capacitance Combinations
for the Remote Capacitor Selector of Fig. 10-20.**

Inputs				Capacitance
A	B	C	D	
0	0	0	0	0.005 μF
0	0	0	1	0.015 μF
0	0	1	0	0.025 μF
0	0	1	1	0.035 μF
0	1	0	0	0.055 μF
0	1	0	1	0.065 μF
0	1	1	0	0.075 μF
0	1	1	1	0.085 μF
1	0	0	0	0.105 μF
1	0	0	1	0.115 μF
1	0	1	0	0.125 μF
1	0	1	1	0.135 μF
1	1	0	0	0.155 μF
1	1	0	1	0.165 μF
1	1	1	0	0.175 μF
1	1	1	1	0.185 μF

PROJECT 53:
Amplifier with remotely controlled gain

This project is a practical application for the remote resistor selector of project 51. In this case, the selected resistor determines
the gain of a small audio amplifier.

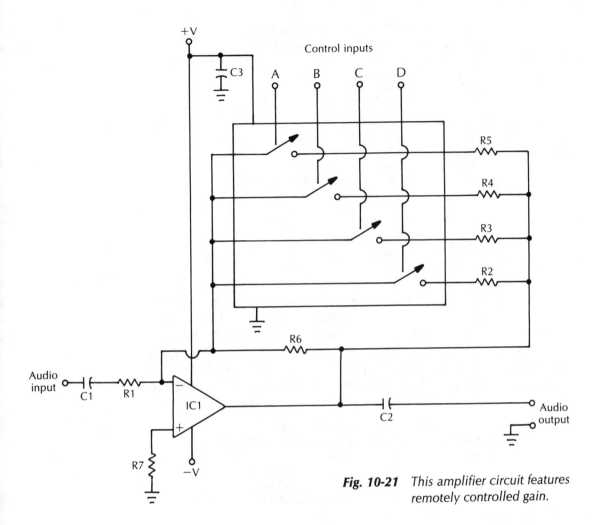

Fig. 10-21 *This amplifier circuit features remotely controlled gain.*

The schematic diagram for this project is shown in Fig. 10-21. The parts list is given in Table 10-10. Almost any standard op amp can be used in this project. While the common and inexpensive 741 will do fine for general purpose testing and low- to medium-fidelity applications, for serious audio applications, a high-grade low-noise op amp IC will do a much better job.

This project is really quite simple. It is simply an op amp used as an audio amplifier in an inverting amplifier configuration. The feedback resistance is controlled by the remote resistor selector. The gain of an inverting amplifier is equal to

$$G = -R_f/R_i$$

Table 10-10   Parts List for the Amplifier
with Remotely Controlled Gain of Fig. 10-21.

IC1	Op amp (741 or better)
C1	0.1-μF capacitor
C2, C3	0.01-μF capacitor
R1	33-kΩ, 0.25-W resistor
R2, R7	47-kΩ, 0.25-W resistor
R3	100-kΩ, 0.25-W resistor
R4	220-kΩ, 0.25-W resistor
R5	330-kΩ, 0.25-W resistor
R6	1-MΩ fixed resistor (experiment)

where

R_f = feedback resistor and
R_i = input resistor.

A large-value fixed resistor (R6) is placed in parallel with the remote resistor selector so the feedback resistance always has a definite, finite value, even when a control signal of 0000 is used.

Each possible gain combination for the resistor values from the parts list is summarized in Table 10-11. Once again, notice that the gain does not increase in linear fashion with increases in

Table 10-11   Available Gains
for the Remotely Controlled Gain Amplifier of Fig. 10-21.

Control inputs				R2	R3	R4	R5	Total feedback resistance	Gain
A	B	C	D						
0	0	0	0	—	—	—	—	1 MΩ	−30.3
0	0	0	1	—	—	—	x	44 kΩ	−1.3
0	0	1	0	—	—	x	—	90 kΩ	−2.7
0	0	1	1	—	—	x	x	30 kΩ	−0.91
0	1	0	0	—	x	—	—	180 kΩ	−5.4
0	1	0	1	—	x	—	x	37 kΩ	−1.1
0	1	1	0	—	x	x	—	64 kΩ	−1.9
0	1	1	1	—	x	x	x	27 kΩ	−0.82
1	0	0	0	x	—	—	—	248 kΩ	−7.5
1	0	0	1	x	—	—	x	39 kΩ	−1.2
1	0	1	0	x	—	x	—	71 kΩ	−2.2
1	0	1	1	x	—	x	x	28 kΩ	−0.85
1	1	0	0	x	x	—	—	116 kΩ	−3.5
1	1	0	1	x	x	—	x	33 kΩ	−1
1	1	1	0	x	x	x	—	53 kΩ	−1.6
1	1	1	1	x	x	x	x	25 kΩ	−0.76

the input control value. Instead, the gain goes up and down in a more or less random pattern. This is because of the way resistances combine in parallel and cannot easily be corrected in a project of this type.

All of the gains are given as negative numbers. This indicates the polarity inversion of the op amp's inverting input. Of course, gain values of less than one represent attenuation of the input signal. By all means, feel free to experiment with other resistor values in this circuit.

PROJECT 54:
Remotely controlled potentiometer

Figure 10-22 shows a different approach to the remote selection of resistance values. Here, an X9MME digitally controlled potentiometer is used. This device was discussed in some detail back in chapter 1.

A typical parts list for this project appears as Table 10-12. Select the specific X9MME device you will use for the maximum resistance suitable to your individual application.

An advantage of this project over the remote resistor selector of project 51 is that fewer control lines are required for this circuit. Only three control lines from the remote station are required: ground, up/down select, and increment. Except for the common ground line, the other two control lines are single-bit digital signals, either high or low, with no other possible values. Another advantage of this project is that the controlled resistance does not jump up and down with consecutive steps of the control signal.

IC2 and its associated components are wired as an astable multivibrator, or system clock. The digital potentiometer's current wiper position (determined by the most recent up/down and increment signals) is stored each time the clock signal goes from low to high. Using the component values from the parts list, the clock's timing period is about 1 second. Of course, you can substitute other timing components to suit your individual application.

This circuit can be used in place of any manual potentiometer or fixed resistor. To use it as a fixed resistor, make the connections to pin 5 (V_w) and either pin 6 (V_l) or pin 3 (V_h). If the V_h connection is used, the resistance will vary in the opposite direc-

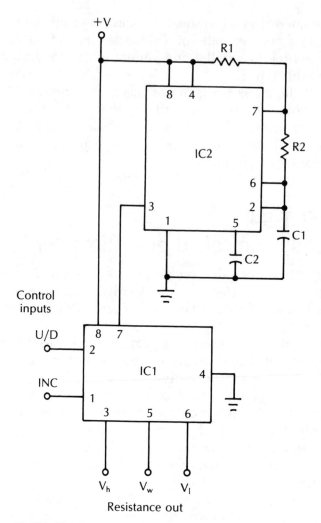

Fig. 10-22 *This is a remotely controlled potentiometer circuit.*

Table 10-12   Parts List for the
Remotely Controlled Potentiometer of Fig. 10-22.

IC1	X9MME digitally controlled potentiometer (X9103 = 10 kΩ; X9503 = 50 kΩ; and X9104 = 100 kΩ)
IC2	555 timer
C1	3.3-μF, 25-V electrolytic capacitor (see text)
C2	0.01-μF capacitor
R1	4.7-kΩ, 0.25-W resistor (see text)
R2	220-kΩ, 0.25-W resistor (see text)

tion as the wiper. That is, when the wiper is moved up, the resistance will decrease, and vice versa. Using the V_1 pin for a two-connection installation causes the resistance to vary directly with the wiper. When the wiper is moved up, the resistance will increase, and vice versa. This is obviously more logical for most applications.

❖ 11
Timers

VIRTUALLY ALL AUTOMATION APPLICATIONS REQUIRE SOME SORT of timer. Sometimes a 24-hour clock will be used to turn something on or off at a specific time. In other cases, some action will be periodically repeated at regular intervals. In still other applications, a timer will be used to introduce a delay between events.

There are many different possible approaches to timer circuitry. IC timer devices are widely available, inexpensive, and easy to use, so there is little point in using discrete components for timing applications. In this chapter, we will look at some of the most popular timer ICs and some of the ways they can be employed in automation applications.

The 555 timer

Without the slightest doubt, the most popular timer device is the 555 IC. The pin diagram for this device is shown in Fig. 11-1. Entire books have been written about just this one device and its many applications.

Simple timers can be built from discrete components such as bipolar and unijunction transistors, but such circuits have some important limitations. For one thing, they tend to be very power supply dependent. If the supply voltage changes for any reason, the timing period can be altered, often by a significant amount.

The 555 timer IC employs a rather tidy solution to this potential problem. The timing period is not dependent on the absolute voltage. Instead, the 555 operates by using the ratio of the supply

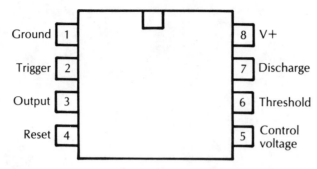

Fig. 11-1   *The 555 is unquestionably the most popular timer IC around today.*

voltage and certain IC terminal voltages. If the supply voltage changes, the terminal voltages will also change by a like amount, holding the ratio constant.

The 555 timer IC is very tolerant when it comes to the power supply. It will work just fine with supply voltages ranging from + 4.5- to + 15-V dc. In addition, although it is principally a linear device, the 555 can easily be directly interfaced to TTL or CMOS digital circuits.

Some of the pin labels may not be familiar to you, so let's take a closer pin-by-pin look at the 555 timer.

- **Pin 1—Ground.** This one is clear enough. It is the ground (common) connection for the supply and signal voltages.
- **Pin 2—Trigger.** This pin is used to initiate a timing cycle. The 555 is triggered by dropping pin 2 below + 1/3 V. Ordinarily, this pin is held at a level above this point. The triggering voltage is typically about one-half the voltage applied to pin 5. Triggering is level sensitive, so slow-changing waveforms (such as sine waves) can be used, as well as more traditional and direct switching waveforms (such as rectangular waves). If the trigger input is held below + 1/3 V for longer than the timing period, the trigger is immediately retriggered. This may or may not be desirable, depending on the application. Generally, the triggering signal should be at least 1 μs (0.000001 second) for reliable triggering.
- **Pin 3—Output.** This pin is also pretty self-explanatory. This is where the timing pulses generated by the 555's internal circuitry can be tapped off. Note that the output of a timer is like a digital signal. It can take on either of two

states: a low level (near ground) or a high level (near V+). There are no intermediate output levels.

- **Pin 4—Reset.** This pin is used to reset the internal latch and drive the output back to its normal low state. The reset threshold level is 0.7 V and a sink current of 0.1 mA is required for resetting the timer. These values are independent of the supply voltage. The reset pin serves an overriding function. It can force the timer to the reset condition, regardless of the signals at any of the other input pins. It is used to tell the timer that the timing period is over.

- **Pin 5—Control voltage.** This pin allows an external voltage to control the switching levels of the internal comparators. This permits a great deal of flexibility in using the timer. Voltage-controlled oscillators and timers and similar applications take advantage of this pin.

- **Pin 6—Threshold.** This pin is one of the inputs to the upper internal comparator. It is used to reset the internal latch (which drives the output low). Resetting with pin 6 is accomplished by raising the voltage on this pin to a level of about +2/3 V. This function is level sensitive, permitting the use of slowly changing waveforms.

- **Pin 7—Discharge.** This is the collector of an internal discharge transistor. The transistor is on when the output is low and is cut off when the output is high. This transistor switch clamps the appropriate nodes of the timing network to ground.

- **Pin 8—V+.** This is the positive supply voltage terminal.

Virtually all timer applications are variations on multivibrators. A multivibrator is a circuit with two possible output states. The output is either at a low level or a high level. There are no intermediate output levels. There are three types of multivibrator circuits.

A *monostable multivibrator* has one stable output state. When triggered, the output goes to the other output state for a specific length of time and then reverts back to the original stable state. The action of a monostable multivibrator is illustrated in Fig. 11-2.

A *bistable multivibrator* has two stable output states. Either output state can be held indefinitely, until a trigger pulse is received, at which point the output reverses states. In a sense, a bistable multivibrator "remembers" its previous output state.

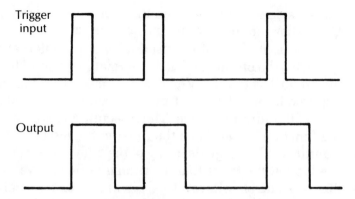

Fig. 11-2 *The monostable multivibrator briefly reverses its output state when a trigger pulse is received.*

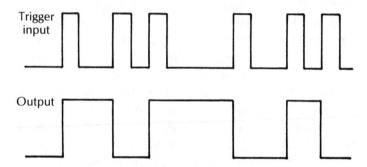

Fig. 11-3 *A bistable multivibrator "remembers" its previous state, reversing output states each time a trigger pulse is received.*

The action of a bistable multivibrator is illustrated in Fig. 11-3.

An *astable multivibrator* has no stable output states at all. It continuously switches back and forth between output states with no trigger pulse input. An astable multivibrator is essentially a rectangular-wave (or square-wave) generator, as shown in Fig. 11-4.

For our purposes, we are concerned with monostable and astable multivibrators. Timer ICs like the 555 can be used in both of these applications.

The basic circuit for a monostable multivibrator built around a 555 timer IC is shown in Fig. 11-5. This is its most basic mode of operation. Notice that there are only three external components: two capacitors and one resistor.

Strictly speaking, capacitor C2 isn't absolutely necessary in many applications. Its purpose is to improve noise immunity.

Fig. 11-4 *An astable multivibrator is free running and does not require a trigger signal to reverse its output states.*

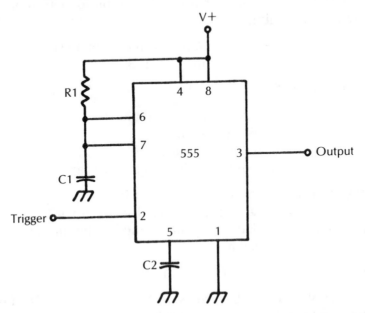

Fig. 11-5 *A simple monostable multivibrator circuit can be built around the 555 timer.*

The 555 does pretty well in this respect on its own. Still, it is a good idea to include this capacitor, just in case. After all, it doesn't cost much and doesn't take up much room. Typically, the value of this capacitor should be between about 0.0005 and 0.05 μF.

The other two components, R1 and C1, define the timing period of the circuit. The formula couldn't be much simpler:

$$T = 1.1\,R1C1$$

where

T = timing period in seconds
R1 = timing resistance in ohms, and
C1 = timing capacitance in farads.

For example, if R1 is 470 kΩ (470,000 Ω), and C1 is 0.22 μF (0.00000022 farad), the timing period will be

$$T = 1.1 \times 470{,}000 \times 0.00000022$$
$$= 0.11374 \text{ seconds}$$

For reliable operation, the timing resistance should be between approximately 10 kΩ (10,000 Ω) and 14 MΩ (14,000,000 Ω), and C1 should be between 100 pF (0.0000000001 F) and 1000 μF (0.001 F). This means that the timing period can range anywhere from a minimum of

$$T = 1.1 \times 10{,}000 \times 0.0000000001$$
$$= 0.0011 \text{ ms} = 1.1 \text{ μs}$$

to a maximum of

$$T = 1.1 \times 14{,}000{,}000 \times 0.001$$
$$= 15{,}400 \text{ seconds}$$
$$= 256.6667 \text{ minutes}$$
$$= 4 \text{ hours, } 16 \text{ minutes, and } 40 \text{ seconds}$$

That is quite an impressive range, isn't it?

In most practical applications, the designer is trying to achieve a timing period of a specific length. The timing equation can be rearranged like this:

$$R1 = T/1.1\, C1$$

A likely value is arbitrarily selected for C1, and this equation is used to solve for R1. If the resulting resistance value is awkward or outside the acceptable range, you can select a new value for C1 and try again.

Let's say we need a timing period of 1 second. We will use a value of 0.1 μF (0.0000001 F) for C1. This means we will need a timing resistance equal to

$$R1 = 1/(1.1 \times 0.0000001)$$
$$= 9{,}090{,}909 \text{ Ω}$$

A 9.1 MΩ (9,100,000 Ω) value is nominally a standard resistance value, and it is very close to the calculated value. But, in actuality, a resistor with this value would be pretty difficult to find. We'd probably do better with a smaller resistance. We therefore need to try another, larger value for C1. This time we will try a 22-μF (0.000022-F) capacitor:

$$R1 = 1/(1.1 \times 0.000022)$$
$$= 41{,}322 \text{ Ω}$$

A 42-kΩ resistor would probably be close enough; or we can use a more readily available 39-kΩ resistor and a 2.2 kΩ resistor in series, giving a total nominal resistance of 4.1 kΩ (41,200 Ω).

There is some leeway in the values because the component tolerances imply that they probably won't be exactly at their nominal values anyway. For precision applications, high-quality (low-tolerance) components should be used. In addition, a variable resistance (potentiometer) can be used to fine tune for the exact timing period desired. A 10-turn potentiometer offers the most precise control.

Now, let's examine how this monostable circuit actually works. As long as a voltage greater than $+ 1/3$ V is applied to the trigger input (pin 2), the timer remains in its standby mode. The output is low, at approximately ground potential (0 V).

If a voltage of less than $+ 1/3$ V (a negative-going pulse) is applied to the trigger input, the timer is turned on and begins its timing cycle. The output snaps high to a level just slightly below $V +$. The capacitor starts to charge up. At some point, the voltage across C1 will exceed $+ 2/3$ V. This voltage is fed to the threshold input (pin 6). As soon as this voltage exceeds $+ 2/3$ V, the timer realizes that the timing period is over and it returns to the

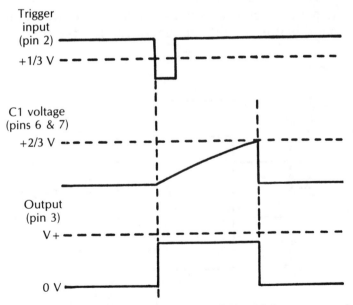

Fig. 11-6 *The key signals from the circuit of Fig. 11-5 are summarized in this timing diagram.*

standby mode. The output snaps back to its low (ground) level. Capacitor C1 discharges through pin 7 of the IC.

The charging rate of the capacitor is determined by the RC time constant of R1 and C1. Obviously, the longer the time constant is, the longer it will take for the capacitor to build up a charge of +2/3 V, and the longer the timing period of the circuit will be.

Figure 11-6 shows a timing diagram summarizing the key signals within this circuit. Notice that all operating voltages are based on the value of the supply voltage (V+). If the supply voltage should change for any reason, the operating voltages will all change by a similar amount, so circuit operation will be virtually unaffected.

Notice also that the length of the triggering input pulse has no effect on the length of the output timing period. (Unless the triggering pulse is longer than the timing period.) Monostable multivibrators are sometimes called pulse stretchers because the output pulse is longer than the input pulse.

The other basic timer application is the astable multivibrator. The basic 555 astable multivibrator circuit is illustrated in Fig. 11-7. Notice how similar it is to the monostable circuit presented earlier in this section.

The main differences are that there are two timing resistors (R1 and R2) in this circuit, and the trigger input (pin 2) is shorted

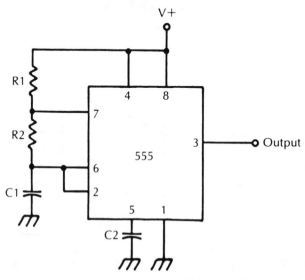

Fig. 11-7 *The other basic application for the 555 timer IC is in astable multivibrator circuits.*

to the threshold input (pin 6). No external input signal is used in the basic astable multivibrator. It is free running.

When power is initially applied to this circuit, the voltage across timing capacitor C1 will, naturally, be low. As a result of this, the timer is triggered (through pin 2). The output goes to its high state, and the internal discharge transistor (at pin 7) is turned off. A complete circuit path through C1, R1, and R2 is formed, charging the capacitor. When the charge on the capacitor exceeds $+2/3$ V, the upper threshold is reached. This voltage on pin 6 forces the output back to its low state.

Timing capacitor C1 now starts to discharge through R2 (but not R1). When the voltage across the capacitor drops below $+1/3$ V, the timer is automatically retriggered and a new cycle begins.

The timing signals within this circuit are illustrated in Fig. 11-8. Notice that since both R1 and R2 affect the charging time (high output—T1), and only R2 affects the discharging time (low output—T2), the low output time will always be at least slightly less than the high output time. A true 50% duty cycle square

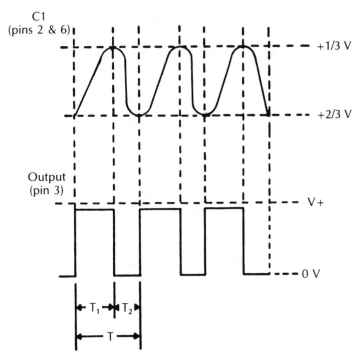

Fig. 11-8 *The key signals from the circuit of Fig. 11-7 are summarized in this timing diagram.*

wave is not possible with this circuit, although you can come pretty close if the value of R1 is relatively small with respect to R2.

To predict the action of this circuit, we need several simple formulae. The charging time (high output time) equals

$$T1 = 0.693(R1 + R2)C1$$

and the discharging time (low output time) equals

$$T2 = 0.693R2C1$$

The total time of the entire cycle is simply the sum of the charge and discharge times:

$$
\begin{aligned}
T &= T1 + T2 \\
&= 0.693(R1 + R2)C1 + 0.693R2C1 \\
&= 0.693(R1 + 2R2)C1
\end{aligned}
$$

The frequency of the output waveform is the reciprocal of the time of the entire cycle. That is

$$F = \frac{1}{T}$$

With a little bit of algebraic rearranging, the frequency formula becomes

$$F = \frac{1.44}{(R1 + 2R2)C1}$$

The same limitations on the component values hold for the monostable multivibrator. The combined sum of R1 and R2 should be between 10 kΩ and 14 MΩ. The timing capacitor (C1) should have a value between about 100 pF and approximately 1,000 μF.

Let's work through a few typical examples. To begin with, let's assume we have the following component values:

- R1 = 47 kΩ (47,000 Ω)
- R2 = 47 kΩ (47,000 Ω)
- C1 = 0.1 μF (0.000001 F).

In this case, the output will be high for a time equal to

$$T1 = 0.693 \ (R1 + R2)C1$$
$$= 0.693 \times (47,000 + 47,000) \times 0.0000001$$
$$= 6.6 \ \text{ms}$$

The low output time will be equal to

$$T2 = 0.693 \ R2C1$$
$$= 0.693 \times 47,000 \times 0.0000001$$
$$= 3.3 \ \text{ms}$$

This means that the entire output cycle will last approximately

$$T = T1 + T2$$
$$= 0.0065 + 0.0033$$
$$= 9.8 \ \text{ms}$$

Taking the reciprocal of this, we get an approximate output frequency of

$$F = \frac{1}{T}$$
$$= \frac{1}{0.0098}$$
$$\approx 102 \ \text{Hz}$$

We can double-check our work by utilizing the alternate frequency equation:

$$F = \frac{1.44}{(R1 + 2R2)C1}$$
$$= \frac{1.44}{(47,000 + 2 \times 47,000) \times 0.0000001}$$
$$= 102 \ \text{Hz}$$

Notice that using identical values for R1 and R2 certainly does not give us a 1:2 duty cycle waveform.

Let's try a second example. This time we will assume the following component values:

- R1 = 22 kΩ (22,000 Ω)
- R2 = 820 kΩ (820,000 Ω)
- C1 = 0.47 μF (0.00000047 F).

The cycle times in this example work out to

- T1 = 0.2742 second = 274.2 ms
- T2 = 0.2671 second = 267.1 ms
- T   = 0.5413 second = 541.3 ms.

Notice that since R2 is so much larger than R1, the result is very close to a 1:2 (50%) duty cycle square wave.

Finally, we can solve for the output frequency of our sample circuit:

$$F = \frac{1.44}{(22{,}000 + 2 \times 820{,}000) \times 0.00000047}$$
$$= \frac{1.44}{(1{,}662{,}000 \times 0.00000047)}$$
$$= \frac{1.44}{0.78114}$$
$$\approx 1.84 \text{ Hz}$$

In many practical applications, we will only be interested in the output frequency, and we don't have to bother with calculating the timing periods at all. Large component values result in low output frequencies. For instance, if

- R1 = 3.3 MΩ (3,300,000 Ω)
- R2 = 10 MΩ (10,000,000 Ω)
- C1 = 1,000 μF (0.001 F).

$$F = \frac{1.44}{(3300000 + 2 \times 1000000) \times 0.001}$$
$$= 0.000062 \text{ Hz} = 0.062 \text{ MHz}$$

This is about one complete cycle every 4.5 hours. This is close to the lower limit for the basic 555 astable multivibrator circuit.

On the other hand, small component values result in high output frequencies. As an example, if

- R1 = 6.8 kΩ (6800 Ω)
- R2 = 3.9 kΩ (3900 Ω)
- C1 = 100 pF (0.0000000001 F).

$$F = \frac{1.44}{(6800 + 2 \times 3900) \times 0.0000000001}$$
$$= 986{,}301 \text{ Hz}$$
$$\approx 986 \text{ kHz}$$

As you can see, this circuit is capable of a very wide range of output frequencies.

For stable operation, however, it is not recommended that this circuit be used at frequencies above about 100 kHz. This is due to internal storage times within the 555 chip itself.

Duty cycles of near 50% to 99% can be obtained with the proper selection of values for R1 and R2. R1 may be as small as 100 Ω or R2/100, whichever is larger.

Multiple 555 ICs

The 555 timer is such a versatile chip that it is often used several times within a single circuit. Several 555s are often cascaded to extend the maximum timing period. This technique will be discussed a little later in this chapter.

When two or more 555 timers are employed in a single circuit, individual 555 timer IC packages can be used, of course. But a more elegant solution is also available. Multiple-timer sections in a single IC package are widely available and permit convenient and compact construction of complex circuits.

The 556 dual-timer IC is almost as popular as the regular 555 timer chip. This device contains two independent 555-like timers in a single 14-pin package. The pin diagram of the 556 dual-timer chip is shown in Fig. 11-9. The two timer sections in this

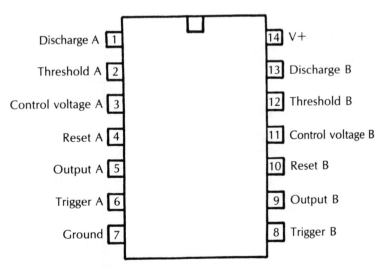

Fig. 11-9 The 556 dual timer IC contains two 555 timers in a single package.

IC are electrically identical to individual 555s in every way. Moreover, the two timer sections are almost completely independent of each other. Only their power supply pins (pin 14—V + — and pin 7—ground) are used in common. All operational pins are fully separate and isolated for the two timer sections. The two sections can be cascaded or used in entirely different portions of the circuit. One section can be used as a monostable multivibrator, while the other is wired as an astable multivibrator, or both sections can be set up as the same type of multivibrator. The timing periods for the two sections are completely independent of each other.

If you need even more timer sections, you might want to try the 558 quad-timer chip. This device is sometimes identified as the 5558. As the pin diagram of Fig. 11-10 indicates, the 558 contains four timer sections that are each similar to the 555, although the available connections are somewhat simplified to limit the number of pins required on the device. The 558 is designed primarily for monostable multivibrator applications, and generally cannot be used in most astable multivibrator circuits.

The timing equations for the 556 and the 558 timer sections are identical to those used for the standard 555 timer. These equations were discussed in detail earlier in this chapter.

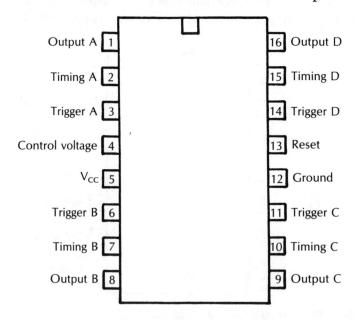

Fig. 11-10 *The 558 is a quad-timer IC.*

Precision timers

The 555 is unquestionably the most popular IC timer. But it is far from the only one available, nor is it the best choice for all applications. The 555 is certainly cheap and convenient, but in some applications, greater precision might be required.

The 322, shown in Fig. 11-11, is a popular precision timer IC. Closely related to the 322 is its "little brother," the 3905, which is illustrated in Fig. 11-12. The internal circuitry of these two devices is essentially the same, so we will concentrate on the 322.

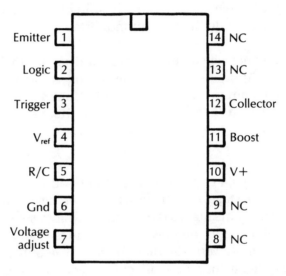

Fig. 11-11   *For precision timer applications, the 322 is often a good choice.*

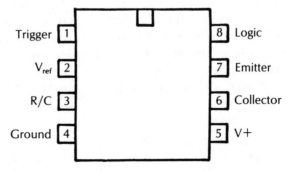

Fig. 11-12   *The 3905 precision timer IC is basically a simplified version of the 322.*

In many ways these chips offer a step up from the standard 555 timer. The most important advantages of the 322 include

- a wider range of acceptable supply voltages
- greater immunity to supply voltage variations
- a wider range of possible timing periods
- simpler calculations
- a higher and more flexible voltage output stage
- improved timing period accuracy.

There is one major disadvantage to the 322 (aside from the higher cost). This device is not really suitable for astable operation. It is only intended for use in monostable multivibrator circuits.

The timing period calculation is simplified for the 322 (compared to the 555) by the fact that the timing voltage reference is 0.632 times the supply voltage. This eliminates the constant in the timing equation. The formula for the timing period of the 322 is simply

$$T = R1C1$$

The timing voltage for the 322 is derived from an internal voltage regulator stage, which provides a precise 3.15 V to the divider network. This makes the circuit even less sensitive to any variations and noise in the power supply lines than the 555.

The 322 is more versatile than the 555. The output stage can be wired in either a common-collector or a common-emitter configuration, depending on the requirements of the specific application at hand. Additional flexibility is offered by the logic pin. In the 555 the output is normally low and goes high only during the timing cycle. On the 322, holding the logic pin high gives the same response. A low signal on the logic pin reverses this action. The output will normally be high and goes low only during the timing cycle. This eliminates the need for a separate inverter stage in many circuits.

Let's take a brief look at each of the pins on the 322 precision timer IC.

- **Pin 1—Emitter (pin 7 on the 3905).** This is an output pin. It is connected to the emitter of the internal NPN transistor. If the output is being taken off of this pin, the collector pin (pin 12) is connected to V+ or some other positive voltage.

- **Pin 2—Logic (pin 8 on the 3905).** This pin is a logic input. It determines the normal output condition. If the logic pin is held low, the output is normally high. If the logic pin is held high, the output is normally low.

- **Pin 3—Trigger (pin 1 on the 3905).** This is the input pin for the trigger pulse that initiates the timing cycle. It is similar to the trigger input on the 555. The timing cycle will not be retriggered if the trigger pulse is longer than the period of the timer. The output will time out and return to its normal state; however, the timing capacitor will not be discharged until the trigger pulse is removed. The trigger input is normally held low. The timer is triggered when this input goes high (positive).

- **Pin 4—V_{ref} (pin 2 on the 3905).** This is a reference voltage output. It is the output of the internal 3.15-V regulator. An external load of up to 5 mA can be applied to this output. The timing resistor (R_t) is normally connected between this pin and the R/C pin (pin 5). This reference voltage is very precise. Drift is typically 0.01%.

- **Pin 5—R/C (pin 3 on the 3905).** This is the common connection point for the timing resistor (R_t) and the timing capacitor (C_t).

- **Pin 6—Ground (pin 4 on the 3905).** This is the ground connection pin for the IC.

- **Pin 7—Voltage adjust (not used on the 3905).** This pin allows the designer to access the comparator reference point for charging and discharging the timing capacitor. The normal comparator reference point is 2 V. If this pin is not used, noise immunity of the circuit can be improved by bypassing it to ground with a capacitor in the 0.01-μF to 0.1-μF range.

- **Pin 8—No connection.**
- **Pin 9—No connection.**
- **Pin 10—V+ (pin 5 on the 3905).** This is the positive supply voltage pin.
- **Pin 11—Boost (not used on the 3905).** Connecting this pin to V+ increases the switching speed of the internal comparator. It is generally used to improve the timing accuracy for periods less than about 1 ms (0.001 second).

- **Pin 12—Collector (pin 6 on the 3905).** This is an alternate output pin. It is the collector of the internal NPN output transistor. If this output is used, the emitter output (pin 1) will generally be grounded. An output from this pin is taken across a load to V+ .
- **Pin 13—No connection.**
- **Pin 14—No connection.**

A basic monostable multivibrator circuit using the 322 precision timer IC is illustrated in Fig. 11-13. The logic input is shown as switchable here. Normally it will be permanently connected to either ground or V_{ref} depending on the desired output pattern.

The value of resistor R1 should be between about 10 kΩ (10,000 Ω) and 100 MΩ (100,000,000 Ω). Capacitor C1 should be between about 100 pF (0.0000001 F) and 1000 μF (0.00 F).

The timing equation is simply

$$T = R1C1$$

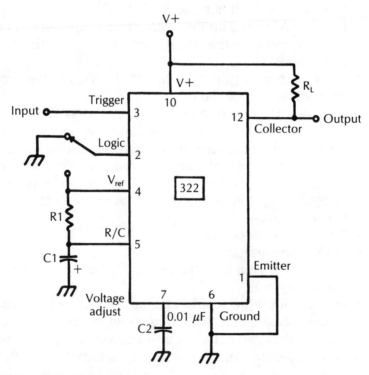

Fig. 11-13   *This is the basic monostable multivibrator circuit using the 322.*

This means the timing period can range from

$$T = 10,000 \times 0.0000000001$$
$$= 0.001 \text{ ms}$$
$$= 1 \text{ } \mu s$$

to

$$T = 100,000,000 \times 0.001$$
$$= 10,000 \text{ seconds}$$
$$= 2 \text{ hours, } 46 \text{ minutes, } 40 \text{ seconds}$$

Certainly an impressive range!

Cascading timers

In many applications, you will need multiple timer circuits. This is often the case when long time delays are required or when multiple events are to occur at varying times.

Timers can be cascaded in parallel or in series. Parallel cascaded timers, as shown in Fig. 11-14, generate multiple outputs that begin together, but can end at different times. A typical timing diagram for this system is shown in Fig. 11-15.

Series cascaded timers, as illustrated in Fig. 11-16, also generate multiple outputs that can be of unequal lengths, but they are sequential rather than simultaneous. The second one begins

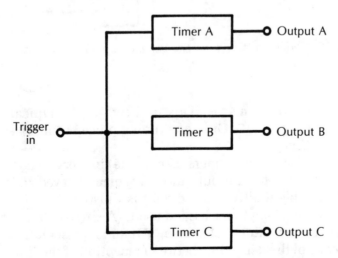

Fig. 11-14 *Parallel cascaded timers generate multiple outputs that begin together but can end at different times.*

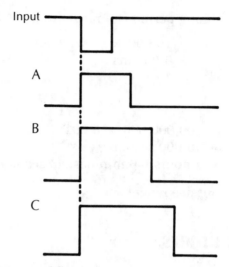

Fig. 11-15 *This is the timing diagram for a typical parallel cascaded timer system.*

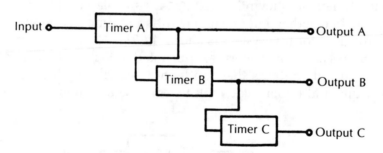

Fig. 11-16 *Series cascaded timers generate multiple output pulses in sequence.*

after the first one finishes. This is shown in the typical timing diagram of Fig. 11-17. Only one timing pulse output is active at any given time.

Series cascading timers can come in very handy for a delayed reaction to an input pulse. A typical delayed action trigger monostable multivibrator circuit is illustrated in Fig. 11-18. Notice that two 555 timer ICs are used. (A single 556 dual-timer IC might be substituted, of course.) IC1 and its associated components control the delay time between reception of the input pulse and the beginning of the output pulse. IC2 and its associated components control the length of the output pulse. A typical

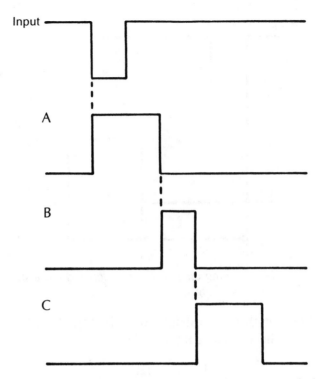

Fig. 11-17   *This is the timing diagram for a typical series cascaded timer system.*

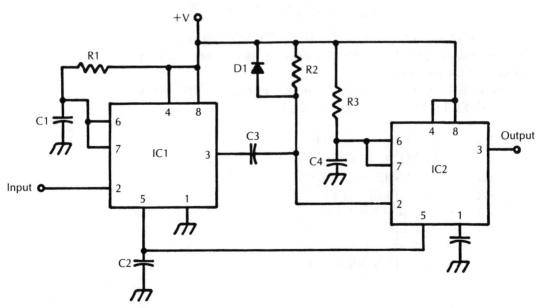

Fig. 11-18   *The output pulse from this monostable multivibrator circuit will be delayed from the trigger pulse.*

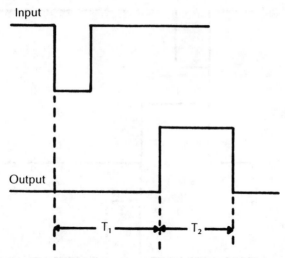

Fig. 11-19 *This is the timing diagram for the delayed monostable multivibrator circuit of Fig. 11-18.*

timing chart for this circuit is shown in Fig. 11-19. The delay time is set by the values of R1 and C1:

$$T_d = 1.1R1C1$$

The output pulse is determined by the values of R3 and C4:

$$T_o = 1.1R3C4$$

Do these equations look familiar? They certainly should. They're just the standard 555 monostable timing equations discussed earlier in this chapter.

We will look at one example of this circuit in use. We will assume the following component values:

- R1　=　470 kΩ (470,000 Ω)
- R2*　=　10 kΩ (10,000 Ω)
- R3　=　220 kΩ (220,000 Ω)
- C1　=　10 μF (0.00001 F)
- C2*　=　0.01 μF (0.00000001 F)
- C3*　=　0.001 μF (0.000000001 F)
- C4　=　0.22 μF (0.00000022 F)
- C5*　=　0.01 μF (0.00000001 F)
- D1*　=　1N914.

The components marked with an asterisk (*) do not affect the output signal and have more or less standardized values. Only

R1, R3, C1, and C4 need to be selected by the designer for the specific application at hand.

Ordinarily, the output of this circuit is low. When an input pulse is received, the output remains low for a delay period equal to

$$T_d = 1.1R1C1$$
$$= 1.1 \times 470{,}000 \times 0.00001$$
$$= 5.17 \text{ seconds}$$

Then the output will go high for a period equal to

$$T_o = 1.1R3C4$$
$$= 1.1 \times 220{,}000 \times 0.00000022$$
$$= 0.05324 \text{ second}$$
$$= 53.24 \text{ ms}$$

The output drops back to its low state 5.22324 seconds after the trigger pulse is first sensed at the input.

The 2240 programmable timer

One of my all-time favorite experimenter chips is an exciting and versatile timing device known as the 2240 programmable timer. This IC has many practical applications, particularly in automation systems. Unfortunately, apparently it is no longer in production. Some 2240s might still be available from surplus dealers. They are certainly worth the extra effort to find them.

But if you can't find any 2240 chips, you aren't completely left in the cold. A little later in this chapter we'll look at how the operation of a 2240 programmable timer can be simulated with more readily available devices, although with less elegance and simplicity. Normally I try to avoid hard to find components in my project books, but this one is useful enough to warrant a brief discussion.

The pin diagram for the 2240 programmable timer is shown in Fig. 11-20. One of the first things you'll probably notice in this diagram is that there are eight timing outputs. These outputs actually come from an internal eight-bit binary counter. By combining these outputs in the proper combinations effective timing periods from 1 to 255 times the base timing cycle period (T) can be created.

The base timing cycle period for the 2240 programmable timer is calculated with the same very simple equation used for

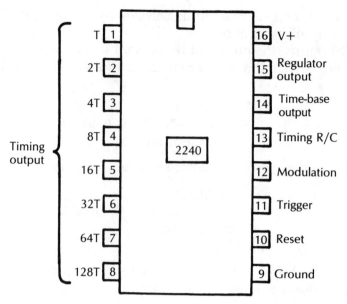

Fig. 11-20 *One of the most versatile timing devices is the 2240 program-mable timer IC.*

the 322 precision timer discussed earlier in this chapter:

$$T = RC$$

where

R = value of an external timing resistor in ohms and
C = value of an external timing capacitor in farads.

The timing period (T) appears at the pin 1 output. The remaining outputs are binary multiples of the basic timing cycle period.

To see how this works, let's try a specific example. We will assume we are using the following component values:

- R = 39 kΩ (39,000 Ω)
- C = 0.5 μF (0.0000005 F).

In this case the base timing cycle period is equal to

$$T = 39,000 \times 0.0000005$$
$$= 0.0195 \text{ second}$$
$$= 19.5 \text{ ms}$$

When the timer circuit is triggered (at pin 11), a 0.0195-second pulse appears at output pin 1. Each of the other seven output pins will generate proportionately longer pulses. For our example, these output timings are equal to

- pin 2: 2T   =   2 × 0.0195 = 0.039 second
- pin 3: 4T   =   4 × 0.0195 = 0.078 second
- pin 4: 8T   =   8 × 0.0195 = 0.156 second
- pin 5: 16T =   16 × 0.0195 = 0.312 second
- pin 6: 32T =   32 × 0.0195 = 0.624 second
- pin 7: 64T =   64 × 0.0195 = 1.248 seconds
- pin 8: 128T = 128 × 0.0195 = 2.496 seconds.

All of these varied timing periods are simultaneously available from a single circuit!

We can also combine two or more output pins to create different, intermediate multiples of the base timing cycle (T). For instance, if we add together pins 1, 2, 4, and 7, the length of the output pulse in our example will be

$$
\begin{aligned}
\text{Output} &= T + 2T + 8T + 64T \\
&= 75T \\
&= 75 \times 0.0195 \\
&= 1.4625 \text{ seconds}
\end{aligned}
$$

The maximum possible timing period for any given 2240 timer circuit will be achieved when all eight of the outputs are summed together. In our example, this works out to

$$
\begin{aligned}
\text{Output} &= T + 2T + 4T + 8T + 16T + 32T + 64T + 128T \\
&= 255T \\
&= 255 \times 0.0195 \\
&= 4.9725 \text{ seconds}
\end{aligned}
$$

In our typical example circuit, we can select output pulses ranging from just under 20 ms up to almost 5 seconds with just a single set of component values!

A broad range of timing component values can be used with the 2240 programmable timer. The timing resistor's value should be between 1 kΩ (1000 Ω) and 10 MΩ (10,000,000 Ω). Similarly, the acceptable range of values for the timing capacitor runs from 0.01 μF (0.00000001 F) up to 1000 ΩF (0.01 F). In addition, the internal circuitry within the 2240 chip itself limits the minimum base timing cycle period to 10 μs (0.0001 second).

Obviously, the maximum base timing cycle period is obtained when the largest acceptable component values are used.

That is, when R equals 10 Ω and C is 1000 μF. In this case, the base timing cycle period works out to

$$T = 10,000,000 \times 0.001$$
$$= 10,000 \text{ seconds}$$
$$= 2 \text{ hours, } 46 \text{ minutes, } 40 \text{ seconds}$$

And since the 2240's programmable outputs can all be selected to give an output pulse of up to 255 times the base timing cycle period, the maximum output pulse is equal to

$$\text{Output} = 255T$$
$$= 255 \times 10,000$$
$$= 2,550,000 \text{ seconds}$$
$$= 708 \text{ hours, } 20 \text{ minutes}$$
$$= 29 \text{ days, } 12 \text{ hours, } 20 \text{ minutes}$$

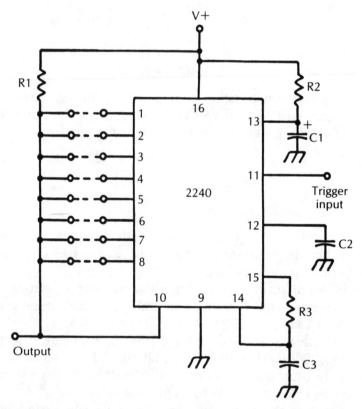

Fig. 11-21 *This is the basic programmable monostable multivibrator circuit using the 2240.*

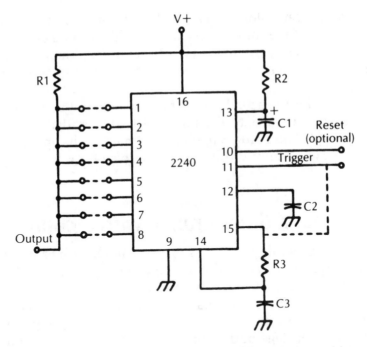

Fig. 11-22   *The 2240 can also be used to build a programmable astable multivibrator circuit.*

That is almost an entire month! The 2240 programmable timer's range should be more than sufficient for the vast majority of practical applications.

Like the simple 555 timer, the 2240 programmable timer can be used in either monostable multivibrator circuits or astable multivibrator circuits. The basic 2240 monostable multivibrator circuit is shown in Fig. 11-21, while the basic 2240 astable multivibrator circuit is illustrated in Fig. 11-22.

In each case, resistor R1 will typically have a value of 10 kΩ (10,000 Ω). This is the load resistor. The output pulses are normally tapped off across this resistor. Resistor R2 is the timing resistor (R) and capacitor C1 is the timing capacitor (C).

Resistor R3 serves as a load resistance for the time-base output. A typical value for this resistor is about 22 kΩ (22,000 Ω). Capacitor C2 is used to improve noise immunity. It typically has a value of about 0.01 μF. It is similar in purpose to the small capacitor connected to the unused voltage control input in most 555 timer circuits. Of course, if an external signal is being fed to the 2240's modulation input (pin 12), capacitor C2 should be eliminated from the circuit.

Finally, capacitor C3 is usually optional. It is generally needed only if the timing capacitor (C1) has a value of 0.1 μF or less and the supply voltage is 7 V or more. When used, this capacitor will usually have a very small value, similar to capacitor C2.

Except for the actual timing components (R2 and C1), none of the exact component values in these basic circuits are particularly critical. The 2240 programmable timer can be operated over a wide range of supply voltages from a low of + 4 V to a high of + 15 V.

Extending a timer's range with a counter

The basic principle of the 2240 programmable timer IC can be used to expand the range of almost any timer chip. The technique is illustrated in block diagram form in Fig. 11-23.

The actual timer must be wired as an astable multivibrator. It's output pulses increment a binary counter one step per pulse. In effect, each stage of the binary counter divides the input frequency by two, doubling the pulse length.

For a long timing period astable multivibrator circuit, the counter should be hooked up to reset itself to zero and automatically repeat the timing cycle when the maximum count is achieved.

For a long timing period monostable multivibrator circuit, the counter should be hooked up to reset itself to zero and stop when the maximum count is achieved. A reset signal is required before the counter will respond to any further pulses from the base timer. This can be a manual reset button or it can be an electrical signal derived from an input trigger pulse.

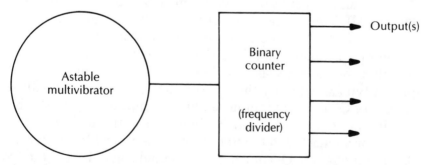

Fig. 11-23 *The operation of a 2240 can be simulated with a standard timer (in the astable mode) followed by a binary counter.*

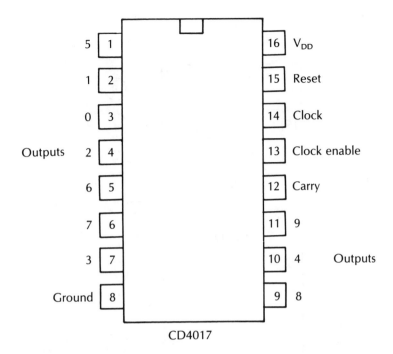

Fig. 11-24 *The CD4017 is an easy-to-use decade counter device.*

A handy counter chip for this type of application is the CD4017 decade counter/decoder. As the pin diagram of Fig. 11-24 shows, this device has ten outputs. At any given instant, one output will be high, while the other nine will be low. Notice that the outputs count decimally, rather than in binary form. This shouldn't matter in most applications. In fact, many users will find this more convenient. Figure 11-25 shows how the CD4017 can be wired to count to N and recycle for astable operation.

For monostable operation, use the count to N and halt the circuit shown in Fig. 11-26. To reset the counter, a brief positive trigger pulse is fed into pin 15.

24-hour clocks

Some automation systems might require a regular 24-hour time-keeping clock, so that events can be triggered at specific times. An electric alarm clock can be easily adapted, simply by tapping off the alarm triggering signal and using it to trigger the automation circuitry. A more inexpensive, compact, and possibly versa-

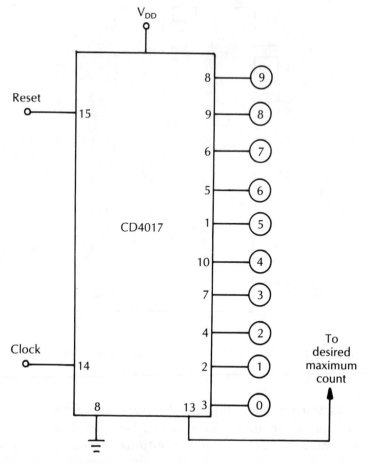

Fig. 11-25   *The CD4017 can be wired to count up to a desired maximum then stop until it is externally reset.*

tile approach might be to use one of the many digital clock ICs on the market. Great bargains are offered by surplus dealers.

Because there are so many clock ICs around, including many discontinued surplus devices, there wouldn't be much point in going into detail about any specific device here. Most clock ICs are very easy to use, as long as you have the manufacturer's specifications sheet at hand.

For less-exacting applications, you can set up an astable multivibrator circuit with a 24-hour cycle period. You may need to cascade several timer stages to achieve this much delay. The 2240 programmable timer IC can generate cycles this long directly by using its upper outputs.

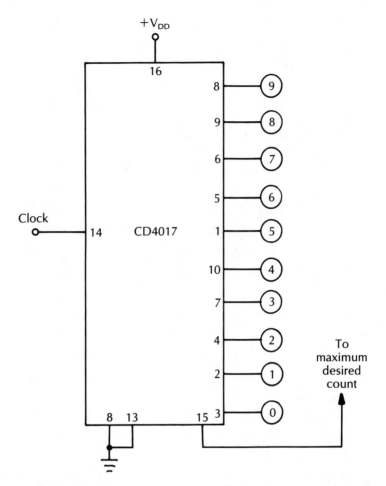

Fig. 11-26 *The CD4017 can also be wired to count up to a desired maxi-mum then automatically reset itself and start over.*

❖ 12
Wireless control

SO FAR, THIS BOOK HAS ASSUMED THAT ALL REMOTE CONTROL signals are transmitted over connecting wires and cables between the controller and the controlled device. This is certainly the simplest and most direct method of communication within the control system. In many applications it will be the best choice, if only for economic reasons. In other applications, however, a connecting cable might be inconvenient or even impossible. Fortunately, other alternatives exist. This chapter explores several popular means of wireless control, including

- light-beam control
- tone control
- carrier-current control
- radio control.

As might well be expected, each of these offers its own set of advantages and disadvantages. There are a number of possible variations within each of these categories.

Light-beam control

Any form of energy can be converted into another form of energy. In our control system, we generally want to start out with electrical energy at the controller and end up with electrical energy at the controlled device. This energy can be converted to another form for transmission.

Light is a form of energy that is relatively easy to convert back and forth from electrical energy. A light bulb or an LED will emit light proportional to the electrical energy being fed to it.

When it comes to converting light back into electrical energy, there are several different approaches. A number of photosensitive devices are available to the experimenter. A photosensitive device is simply one that responds to light. Photosensitive electrical components are often called photocells, but this usage can be rather confusing. Properly speaking, a photocell is a photovoltaic cell or solar cell.

Photovoltaic cells

Photovoltaic cells are basically just a simple PN junction, usually made of silicon. They are essentially similar in construction to standard diodes, but the silicon is exposed to light. Silicon is a photosensitive material. This is why ordinary semiconductor devices are enclosed in a lighttight housing.

Usually the silicon of a photovoltaic cell is spread out into a relatively large, thin plate for the largest possible contact area (exposure to the light source). The symbol for a photovoltaic cell is shown in Fig. 12-1. Notice that it is very similar to the symbol for an ordinary battery (dc voltage source).

When the silicon surface is shielded from light, no current flows through the cell. But when it is exposed to a bright light, a small voltage is generated within the cell because of the photoelectric effect.

If an illuminated photovoltaic cell is hooked to a load, current flows through the circuit, as illustrated in Fig. 12-2. Just how much current flows depends on the amount of light striking the photosensitive surface of the cell. The brighter the light, the higher the current available from the cell.

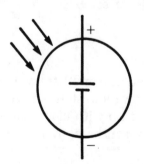

Fig. 12-1 A photovoltaic cell functions as a light-powered "battery."

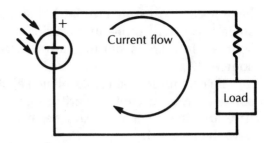

Fig. 12-2 *If an illuminated photovoltaic cell is connected to a load, current will flow through the circuit.*

The photovoltaic's output voltage, on the other hand, is relatively independent of the light level. The voltage produced by most commercially available photovoltaic cells is approximately 0.5 V.

The most obvious use for a photovoltaic cell is as a substitute for an ordinary dry cell. Of course, the 0.5 V output of a single cell is too low for most practical applications, so a number of photovoltaic cells are usually added together in series to form a battery, just as with ordinary dry cells. If the circuit you want to power from photovoltaic cells requires more power than your photocells can provide, more cells can be added in a parallel battery configuration.

You can power almost any dc circuit with a combination of series and parallel connected photovoltaic cells. This is a simple form of solar power, and the combination of cells is called a solar battery, even though it will work just as well under artificial light.

There's one important factor that should be kept in mind— the more cells there are in a solar battery, the larger the total surface area must be and the harder it will be to arrange the cells so they will all be lighted evenly. This means that generally, solar

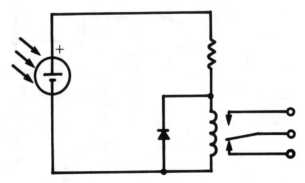

Fig. 12-3 *A photovoltaic cell can be used to trigger a small relay.*

batteries are best suited for fairly low-power circuits. This is why solar power will probably never be a significant primary power source, despite its definite advantages as a secondary power source.

Bear in mind that photovoltaic cells, like any other dc voltage source, have a definite polarity; that is, one lead is always positive and the other is always negative. The connections should never be reversed.

In a control system, a photovoltaic cell can be used to trigger a relay, as illustrated in Fig. 12-3. Note that the controlled circuit requires a separate power supply. The photocell only opens and closes the relay contacts.

Because the output of a photovoltaic cell is fairly small, it can only be used to drive a relatively light-duty relay. If the device you want to control requires a heavier relay, there are two ways to solve the problem. One method is to use the light-duty relay to control the heavy-duty relay, as shown in Fig. 12-4. Alternatively, the output of the photovoltaic cell can be amplified with a transistor, as illustrated in Fig. 12-5. The potentiometer is used to adjust sensitivity (the amount of light required to trigger the relays).

Notice that both of these methods require an extra voltage source in addition to the photovoltaic cell and the controlled cir-

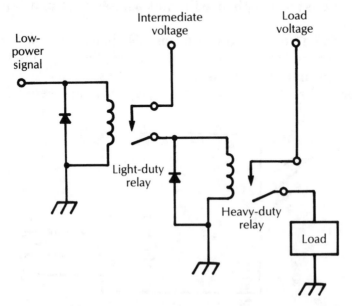

Fig. 12-4 *A light-duty relay can control a larger, heavy-duty relay.*

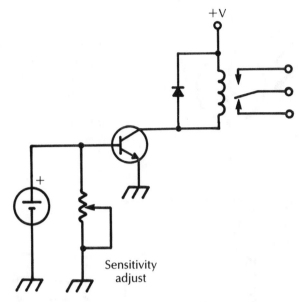

Fig. 12-5 *A larger relay can be driven by amplifying the output of a photo-voltaic cell with a transistor.*

cuit's power supply. All of these relay circuits respond only to the presence or absence of a given amount of light. Because a photovoltaic cell's voltage output is relatively constant, this device cannot be used to monitor varying light levels.

Photoresistors

Another popular light-sensitive device is the photoresistor, or light-dependent resistor (LDR). As the name implies, a photoresistor changes its resistance value in step with the level of illumination on its surface. Photoresistors generate no voltage themselves. They are usually made of cadmium sulfide or cadmium selenide.

These devices can generally cover quite a broad resistance range—often on the order of 10,000:1. Maximum resistance (typically about 1 MΩ (1,000,000 Ω) is usually achieved when the cell is completely darkened. As the light level increases, the resistance decreases.

Photoresistors are junctionless devices just like regular resistors and they have no fixed polarity. In other words, they can be hooked up in either direction without affecting circuit operation. The symbol for a photoresistor is shown in Fig. 12-6.

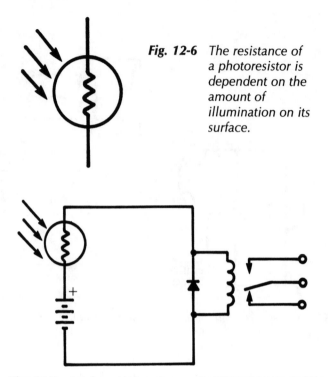

Fig. 12-6 *The resistance of a photoresistor is dependent on the amount of illumination on its surface.*

Fig. 12-7 *A relay can be triggered with a photoresistor.*

Photoresistors are perfect for a wide range of electronic control applications. They can easily be used to replace almost any variable resistor in virtually any circuit. Photoresistors can also be used in many of the same applications as photovoltaic cells, usually with just the addition of a battery and another resistor. They offer the advantage of being sensitive to different light levels. For example, Fig. 12-7 shows a light-controlled relay. A potentiometer can be added so that the relay switches at any desired light level.

Other photosensitive devices

There are numerous other light-sensitive devices available today. Generally these are photosensitive versions of more familiar semiconductor devices, such as light-activated SCRs (LASCRs), photodiodes, and phototransistors. The symbols for these devices are shown in Fig. 12-8.

Phototransistors are especially useful for a number of applications because they can be used as amplifiers whose effective gain is controlled by light intensity. Usually, on phototransistors,

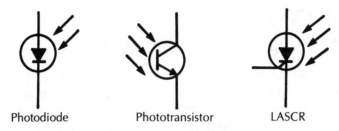

Photodiode   Phototransistor   LASCR

Fig. 12-8 *Light-controlled versions of several standard semiconductor devices are available, including photodiodes, phototransistors, and LASCRs.*

the actual base lead is left unconnected in the circuit. The base-collector current is produced by the photoelectric effect.

Many circuits that use bipolar transistors can be rebuilt with phototransistors, resulting in some very unique effects. Of course, the base lead can be used too. In this case the output will depend on both the signal on the base lead and the light intensity.

An optoisolator is another extremely useful device. As the name implies, it isolates two interconnected circuits so that their only connection is optical (light). Most practical optoisolators consist of an LED and a photoresistor, photodiode, or phototransistor encapsulated in a single lighttight package. The symbol for a typical optoisolator is shown in Fig. 12-9.

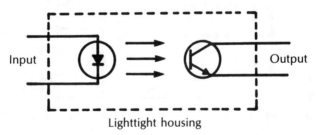

Input   Output

Lighttight housing

Fig. 12-9 *An optoisolator contains a light source and a light detector in a single lighttight housing.*

The LED is wired into the controlling circuit and the phototransistor (or other photosensitive device) is wired into the circuit to be controlled. This provides a convenient means of control with virtually no undesirable cross-talk between the two circuits. Essentially the same effects can be achieved with a separate LED and phototransistor (or photoresistor), but they must be carefully shielded from all external light to prevent any uncontrolled interference.

PROJECT 55:
Latching light-controlled relay

Now that we have gained some familiarity with light-sensitive devices, we can put some of them to work in a few practical control projects. A practical light-controlled relay circuit is illustrated in Fig. 12-10. The parts list is given in Table 12-1.

A latching relay should be used in this project. The relay is activated by momentarily shining a flashlight (or other convenient light source) on the photoresistor. The resistance drops as

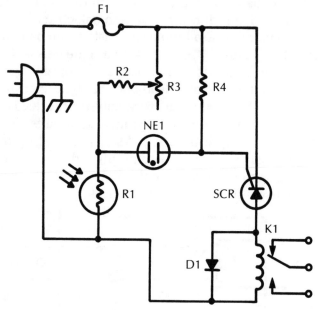

Fig. 12-10   *This is a practical light-controlled relay circuit.*

**Table 12-1   Parts List for the
Latching Light-Controlled Relay of Fig. 12-10.**

SCR	200-V, 4-A SCR
D1	1N4002
K1	120-V ac latching relay (see text)
NE1	Neon lamp
F1	4.5-A fuse and holder
R1	Photoresistor
R2	22-kΩ resistor
R3	1-MΩ trimpot
R4	100-Ω resistor

the cell's illumination increases. This permits the voltage across the neon lamp (NE1). The lamp fires, triggering the SCR. The relay's coil is connected across the SCR's output.

By using a latching relay, the light can be removed from the photoresistor without releasing the relay's switching contacts. The controlled device can be turned off by shining the flashlight on the photoresistor a second time.

PROJECT 56:
Adjustable light-controlled relay

One problem with the last project is that the light level required to trigger the relay is fixed. There is no way for the user to adjust it.

The circuit shown in Fig. 12-11 offers the advantage of being adjustable. The parts list for this project is given in Table 12-2.

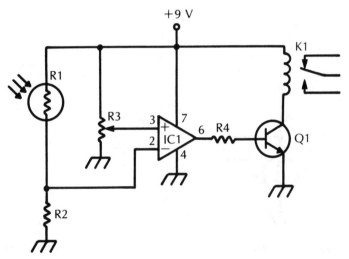

Fig. 12-11   *This light-controlled relay circuit is fully adjustable.*

**Table 12-2   Parts List for the
Adjustable Light-Controlled Relay of Fig. 12-11.**

IC1	Op amp (741 or similar)
Q1	NPN transistor (2N2222 or similar)
K1	9-V relay (contacts to suit application)
R1	Photoresistor
R2, R4	100-kΩ resistor
R3	250-kΩ potentiometer

The op amp is set up as a comparator. The voltage dropped across the photoresistor is compared to the voltage dropped across the upper half of the potentiometer. When the comparator switches over, it triggers the relay. Once again, a latching relay would probably be your best choice for most applications. One light pulse turns the controlled device on. A second light pulse turns it off. The transistor is used to amplify the comparator's output signal. A low-power relay is not needed.

PROJECT 57:
Another light-controlled relay

One more approach to light control of a relay is illustrated in Fig. 12-12. The parts list is given in Table 12-3. When the light level shining on the photoresistor exceeds a specific level, the 555 timer is triggered, activating the relay.

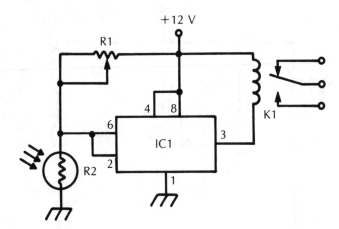

Fig. 12-12 *A different approach to a light-controlled relay is shown here.*

Table 12-3   Parts List for the Light-Controlled Relay of Fig. 12-12.

IC1	555 timer
K1	12-V relay (contacts to suit application)
R1	10-kΩ potentiometer
R2	Photoresistor

PROJECT 58:
Light interruption detector

The circuits discussed so far in this chapter are activated by the presence of light. The circuit shown in Fig. 12-13 is used to detect the absence of light. The parts list for this project is given in Table 12-4.

A light source is positioned so that it shines continuously on the phototransistor. If an object passes between the light source and the phototransistor, the light beam is cut off or interrupted. The output of this circuit puts out a pulse.

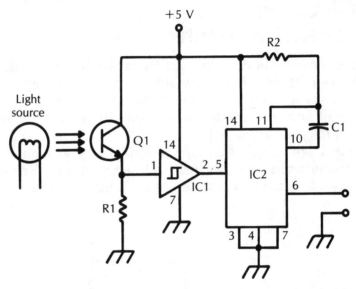

Fig. 12-13 *Sometimes a circuit that detects the absence (instead of the presence) of light can be useful.*

Table 12-4   Parts List for the Light-Interruption Detector of Fig. 12-13.

IC1	7417
IC2	74121
Q1	Phototransistor (TL81 or similar)
C1	0.01-μF capacitor
R1	100-Ω resistor
R2	15-kΩ resistor

The output pulse can be used to trigger a relay or other switching circuit. Another application would be to use this circuit to drive a digital counter. The number of times the light beam is interrupted can be counted, triggering some event when a given number has been passed.

Still another possibility would be to incorporate a timer into the system. Whether or not the light beam is interrupted x number of times within a given period will determine the event controlled by the output. Many other potential applications can be devised for this circuit. As always, use your imagination.

PROJECT 59:
Light-beam transmitter

More advanced applications can be served by modulating the light beam. Modulation is simply the process of imposing an audio (usually) signal onto the carrier (light beam). The light beam will dim and brighten in step with the amplitude fluctuations of the program signal.

You might expect this type of application to require some very complex circuitry, but that isn't necessarily so. The circuit illustrated in Fig. 12-14 is quite simple. It includes just a handful

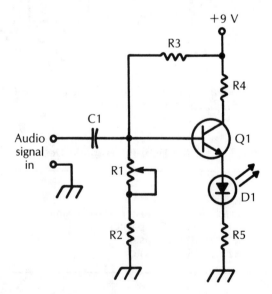

Fig. 12-14 *More complex control signals can be sent out with a light-beam modulator/transmitter.*

Table 12-5   Parts List for the Light-Beam Transmitter of Fig. 12-14.

Q1	NPN transistor (2N2222 or similar)
D1	LED
C1	0.01-μF capacitor
R1	10-kΩ potentiometer
R2	1.2-kΩ resistor
R3	47-kΩ resistor
R4	2.2-kΩ resistor (see text)
R5	15-Ω resistor

of components, but it will do an acceptable job as a light-beam modulator/transmitter.

With no input signal, potentiometer R1 is adjusted until the LED is glowing at half brightness. Now, when an audio signal is applied to the input, the LED's brightness will vary above and below the midpoint (zero level).

Resistor R4 limits the current flow through the LED. You might want to experiment with different values to optimize the performance for the specific transistor and LED you are using.

The modulated signal can be monitored with an ac voltmeter or an oscilloscope at the junction of the LED and resistor R5. The parts list for this project is given in Table 12-5.

PROJECT 60:
Light-beam receiver

Of course the transmitter in project 59 won't be of much use without a suitable receiver. A light beam receiver/demodulator circuit is shown in Fig. 12-15. The parts list is given in Table 12-6. It is designed to complement the transmitter circuit of Fig. 12-14.

The two circuits must be positioned so that the light from the LED in the transmitter reaches the solar cell of the receiver. The range of these circuits is rather limited, but it will be adequate for many remote control applications. The range will be better in a darkened environment. Obviously, any external light source will interfere with the light from the LED. A shield over the photosensor will help some. An infrared emitter and sensor will do somewhat better than if visible light is used.

If the wireless function is not absolutely essential, a fiberoptic connection can be used, greatly increasing the range and per-

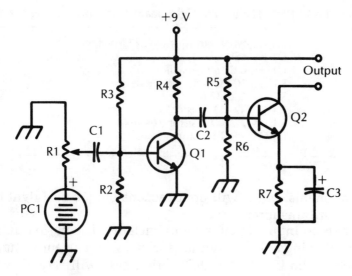

Fig. 12-15   *This receiver circuit is intended to pick up the signals sent out by the transmitter of Fig. 12-14.*

Table 12-6   Parts List for the Light-Beam Receiver of Fig. 12-15.

Q1, Q2	NPN transistor (2N4123 or similar)
PC1	Photocell
C1, C2	0.1-μF capacitor
C3	100-μF, 35-V electrolytic capacitor
R1	1-kΩ potentiometer
R2, R4, R5 '	10-kΩ resistor
R3	100-kΩ resistor
R6, R7	1-kΩ resistor

mitting the signal to travel around corners. Fiberoptics will be discussed a little later in this chapter. Potentiometer R1 in the receiver circuit is a gain control.

The output impedance from this circuit is in the moderate range—about 1000 to 3000 Ω. If the device being driven has a different input impedance, you should use a transformer or some other impedance matching circuit.

PROJECT 61:
Infrared transmitter

Using a light-beam transmitter/receiver system like in the last two projects has certain limitations. Primarily, there can be interfer-

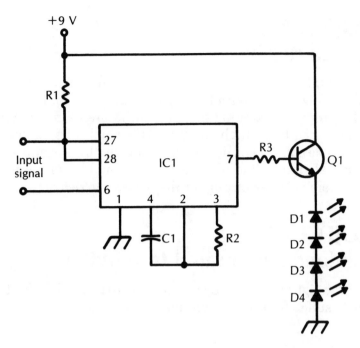

Fig. 12-16 *An infrared beam is invisible to the human eye, but can be transmitted like a regular light beam.*

Table 12-7   Parts List for the Infrared Transmitter of Fig. 12-16.

IC1	ED-15
Q1	NPN transistor (2N4401 or similar)
D1 – D4	Infrared LED (Monsanto MV5000 or similar)
C1	250-pF capacitor
R1	820-Ω resistor
R2	91-kΩ resistor
R3	3.3-kΩ resistor

ence problems from external light sources. Also, in some applications, a visible beam of light might be undesirable.

One solution is to use an infrared light beam, which is invisible to the human eye. An infrared transmitter circuit is illustrated in Fig. 12-16. The parts list is given in Table 12-7.

This transmitter circuit operates at approximately 25 kHz. The multiple infrared LEDs in series increase the range of the system.

PROJECT 62:
Infrared receiver

A compatible infrared receiver to go with the last project is shown in Fig. 12-17, with the parts list given in Table 12-8.

By properly arranging the array of infrared sensors, the system range can be as high as 30 ft. This should be adequate for most remote control applications. Of course, this system can only function along the line of sight. If anything blocks the infrared beam, there will be nothing for the receiver to detect.

PROJECT 63:
Multifunction infrared transmitter

All of the light-control projects we have discussed so far in this chapter suffer from one important limitation, each circuit can

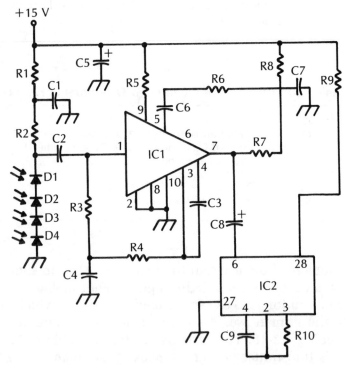

Fig. 12-17　This is the infrared receiver circuit for the transmitter circuit of Fig. 12-16.

Table 12-8   Parts List for the Infrared Receiver of Fig. 12-17.

IC1	CA3035
IC2	ED-15
D1 – D4	Infrared sensor (BP104 or similar)
C1	0.01-μF capacitor
C2	4700-pF capacitor
C3	0.033-μF capacitor
C4	0.025-μF capacitor
C5	10-μF, 35-V electrolytic capacitor
C6	0.0025-μF capacitor
C7	0.05-μF capacitor
C8	2.2-μF, 35-V electrolytic capacitor
C9	270-pF capacitor
R1	100-kΩ resistor
R2, R6	10-kΩ resistor
R3	33-kΩ resistor
R4	150-kΩ resistor
R5	1-kΩ resistor
R7	100-Ω resistor
R8	4.7-kΩ resistor
R9	2.7-kΩ resistor
R10	91-kΩ resistor

control only a single function. Of course, this is just fine for turning a single device on and off. However, many practical remote control applications are more complex. For example, a TV remote control should allow you to turn the set on and off, adjust the volume, and change the channel. Obviously, multiple channel signals are required.

You could build several separate single-channel control devices, but this would rapidly become unwieldy and expensive. We need a true multifunction transmitter/receiver combination.

In most cases, a digital circuit is your best choice because of the convenience of digital gating for separating the encoded signals. A multifunction infrared transmitter circuit is illustrated in Fig. 12-18, with the parts list given in Table 12-9.

Sixteen NO (normally open) SPST switches arranged in an x/y matrix are used to manually enter data. A 16-key keypad would probably be the most convenient choice. Up to 15 functions can be controlled from these switches. (The ''0'' key is ignored.)

IC1 is an encoder that converts the appropriate switch number into a four-bit digital value:

0	0000
1	0001

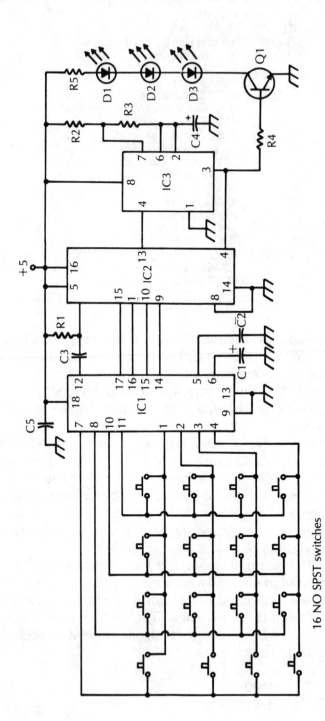

16 NO SPST switches

Fig. 12-18 *Up to 15 independent control signals can be sent out with this infrared transmitter circuit.*

2	0010
3	0011
4	0100
5	0101
6	0110
7	0111
8	1000
9	1001
A(10)	1010
B(11)	1011
C(12)	1100
D(13)	1101
E(14)	1110
F(15)	1111

This binary value is fed to the data inputs of the counter (IC2). The borrow output (pin 13) goes high and activates the multivibrator circuit built around the 555 timer (IC3). The 555's output is connected to the count down input (pin 4) of the counter (IC2). The counter counts down from the loaded data value to zero. (That is why the ''0'' key doesn't work—there aren't any counting pulses.)

On each counting pulse the transistor is turned on, causing the infrared LEDs (LED1–LED3) to flash. Multiple LEDs are used to increase the output level of the infrared signal, maximizing the transmission range. The range is about 15 ft.

Table 12-9   Parts List for the Multifunction Infrared Receiver of Fig. 12-19.

IC1	74922 sixteen-key keyboard encoder
IC2	74193 synchronous up/down dual clock counter
IC3	555 timer
Q1	NPN transistor (2N2222 or similar)
D1–D3	Infrared LED (XC880 or similar)
C1	1-μF, 25-V electrolytic capacitor
C2, C5	0.1-μF capacitor
C3	0.001-μF capacitor
C4	2.2-μF, 25-V electrolytic capacitor
R1	12-kΩ resistor
R2	1-kΩ resistor
R3	1.8-kΩ resistor
R4	2.2-kΩ resistor
R5	100-Ω resistor

The LEDs flash the number of times equal to the value of the closed switch. When the counter reaches zero, it cuts off the timer and waits for the next key to be depressed. All of this happens much faster than you'll be able to notice, even if you could see the infrared pulses. Even the highest number of pulses (15) will be completed in a tiny fraction of a second. You won't have to wait at all before pressing the next key. Your reflexes can't out-race the counter.

You should be careful not to depress two or more keys at once, however. This can confuse the circuit and the output will be unpredictable.

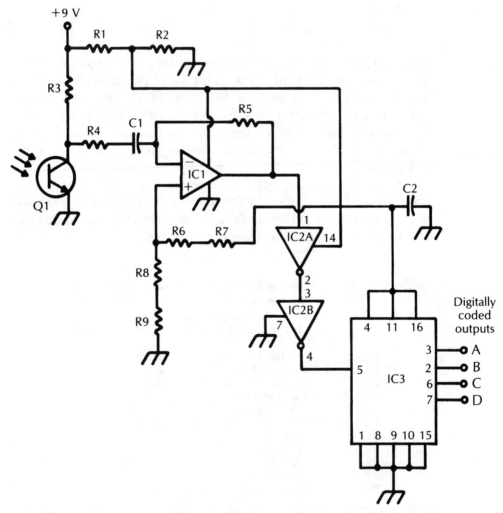

Fig. 12-19 *Here is a receiver for the transmitter circuit of Fig. 12-18.*

PROJECT 64:
Multifunction infrared receiver

Figure 12-19 shows a receiver circuit for use with the transmitter of Fig. 12-18. The parts list is given in Table 12-10.

**Table 12-10 Parts List for the
Multifunction Infrared Receiver of Fig. 12-19.**

IC1	Op amp (748 or similar)
IC2	7414 hex inverter
IC3	74193 synchronous up/down dual-clock counter
Q1	Phototransistor (TIL414 or similar)
C1	2500-pF capacitor
C2	0.1-μF capacitor
R1	12-kΩ resistor
R2, R6	5.1-kΩ resistor
R3	33-kΩ resistor
R4	1-kΩ resistor
R5	1-MΩ resistor
R7	470-Ω resistor
R8	270-Ω resistor
R9	3.3-kΩ resistor

The phototransistor (Q1) detects the infrared pulses. The pulses are then boosted by a high-gain amplifier stage (IC1). The pulses are shaped up into nice clean square waves by IC5. The squared pulses are fed to another counter (IC3). The counter counts the pulses and feeds the value in binary form through outputs A, B, C, and D. The digital value can then be fed to an appropriate gating circuit, so that each possible value can drive a different function.

PROJECT 65:
"Chopped" light-activated relay

Several of the earlier projects in this chapter are designed to perform a switching operation in response to the presence or absence of light. In each case, we have assumed that the activating light is in the form of a continuous beam. This is fine for many applications, but it can be problematic in certain specialized applications.

In some circuits it is preferable to use chopped light. A chopped signal is basically just a stream of regular pulses. Therefore, a chopped light is not a continuous beam, but it is modulated (switched on and off) at a regular rate. In most practical applications, the chop frequency (rate at which the light is switched on and off) will be in the mid-audio range, typically between about 400 Hz and 1 kHz (1000 Hz).

A control circuit for a relay operated by a chopped light source is shown in Fig. 12-20. The parts list for this project is given as Table 12-11.

The relay (K1) should be selected to suit the desired load cir-

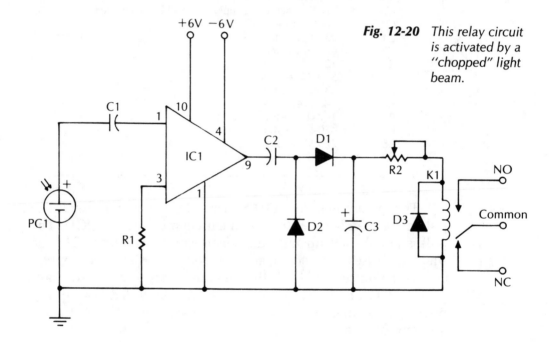

Fig. 12-20 *This relay circuit is activated by a "chopped" light beam.*

Table 12-11   Parts List for the "Chopped" Light-Activated Relay of Fig. 12-20.

IC1	Op amp (CA3010 or similar)
D1, D2	Diode (1N34A or similar)
D3	Diode (1N4002 or similar)
PC1	Photovoltaic cell
C1	15-μF, 15-V electrolytic capacitor
C2	1-μF, 15-V electrolytic capacitor
C3	2-μF, 15-V electrolytic capacitor (see text)
R1	2.7-kΩ, 0.25-W resistor
R2	10-kΩ potentiometer (wire-wound type recommended)

cuit, of course. Diode D3 is included simply to protect the relay's coil from self-destructing due to back-EMF effects.

Potentiometer R2 serves as a sensitivity control for this circuit. Its resistance determines the light intensity required to activate the relay. Depending on your specific application, this might be a front-panel control or a screwdriver-adjusted trimpot. In a few applications, you might prefer to simply replace this potentiometer with a suitably valued fixed resistor. Notice that the op amp (IC1) used in this project requires a dual-polarity power supply. A ± 6-V supply is recommended.

If the controlling light beam's modulation frequency (chop rate) is too low, there might be a slight problem with relay "chatter." That is, the relay's contacts might erratically open and close multiple times instead of neatly making or breaking contact. If you should run into such problems, you can improve the circuit's performance by increasing the value of capacitor C3 slightly.

PROJECT 66:
Light-controlled capacitance

Many electrical parameters can be controlled by a variable resistance. In such cases we can use a photoresistor to place that

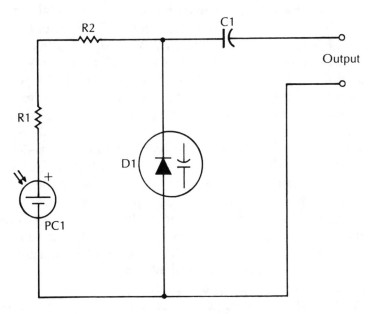

Fig. 12-21 *This is a light-controlled capacitance circuit.*

**Table 12-12 Parts List for the
Light-Controlled Capacitance Circuit of Fig. 12-21.**

D1	Varactor diode (1N4815A or similar) (see text)
PC1	Photovoltaic cell (see text)
C1	0.022-μF capacitor
R1	2.2-MΩ, 0.25-W resistor
R2	3.3-MΩ, 0.25-W resistor

parameter under light control. In some applications, however, we might need a variable capacitance instead of a variable resistance. There aren't any photocapacitors on the market (at least not that I'm aware of). But there's always a solution.

The circuit shown in Fig. 12-21 can be used to control a capacitance via light intensity. The parts list for this project appears as Table 12-12.

A photovoltaic cell (PC1) is used as the light sensor in this project. The exact specifications of this project are not particularly critical. A capacitance that varies in response to the light intensity shining on this sensor will appear across the circuit's output terminals.

The heart of this circuit is the varactor diode (D1). Many electronics hobbyists may not be familiar with varactor diodes, so it might be helpful to briefly describe this unusual type of component here. Basically, a varactor diode is a special two-lead semiconductor that produces an electrically variable capacitance. Actually, any semiconductor diode exhibits some amount of capacitance when it is reverse biased. In the vast majority of standard diode applications, this capacitance is probably undesirable because it can limit the operating or switching frequency of the diode. This can be especially critical in detector or rectifier applications.

A varactor diode, on the other hand, is specially designed to turn this ordinarily undesirable capacitance into a desired and controllable characteristic. The voltage applied to the reverse-biased varactor diode determines its capacitance.

Common applications for varactor diodes include frequency modulation (FM) of oscillators, frequency multipliers, and multiwavelength tuners for RF amplifiers. Because the varactor diode behaves as an electrically controllable capacitance, this type of component is often referred to as a varicap.

A large-value isolating resistor is usually required in varactor diode applications. This is the function of resistors R1 and R2 in this project. Two series resistors are used here for convenience. They are easier to find than a single higher-valued unit. In addition, very high-value resistors can exhibit some stability problems.

Capacitor C1 blocks any dc voltage from the photocell from reaching the circuit's output terminals. This capacitor also makes it impossible for the external load circuit to place a harmful short circuit across the varactor diode (D1). Capacitor C1's value is considerably higher than the maximum varactor capacitance, so it can be reasonably ignored.

The exact range of output capacitances available from this circuit will depend on the characteristics of the specific photocell (PC1) and varactor diode (D1) used. The output capacitance will always be fairly low in value. Typically the range of output capacitances from this type of circuit can be expected to run from a little over 100 to 250 pF. This light-controlled capacitance circuit can be used anywhere you might otherwise use a small-valued variable capacitor.

Fiberoptics

All light-based remote control systems suffer from relatively limited transmission ranges. Moreover, they are, by definition, strictly line-of-sight systems. Light beams cannot travel around corners. You could set up mirrors, but alignment tends to be quite difficult and fussy. This is generally impractical for most control systems. Besides, a person or object could come in between the light source and a critical mirror. Naturally, if any object (including a person or the family cat) gets between the light transmitter and its receiver, the signal will be blocked.

Often the best solution to such problems is to use a fiberoptic cable. In a fiberoptic system, special connecting cables are strung from the transmitter to the receiver, but unlike ordinary wires, fiberoptic cables carry light pulses instead of electrical voltages.

Basically, a fiberoptic cable is a small slender tube made up of multiple hair-thin glass or plastic fibers. Such a cable is quite flexible and can be easily bent in any way needed to conduct the light signal to its desired destination. The light pulses will follow all of the twists and turns of the fiberoptic cable, even if it is tied in a knot. The light at the output is unchanged (except for possible attenuation due to the length of the cable) from the input.

Obviously, if you use a fiberoptic cable, you lose the wireless advantage. So why bother? Fiberoptic cables offer several advantages over ordinary electrical wiring. Because no electrical signal is carried through the cable, it is inherently safer. There is no way for anyone to get shocked or for a fire to start because of worn insulation or a short.

A fiberoptic cable of a given thickness can carry more independent signals than a comparably sized electrical cable. Telephone lines, for example, are using more and more fiberoptics.

Fiberoptic cables tend to be more flexible than comparable electrical cables. If a bend in an electrical cable is too sharp, an internal break in a wire can result.

Fiberoptic cable is more expensive than ordinary electrical wire, so it is only practical for the hobbyist to use it in applications in which its special advantages are significant.

PROJECT 67:
Fiberoptic transmitter

A practical application for fiberoptics is illustrated in Fig. 12-22. This is a remote control transmitter specifically designed for

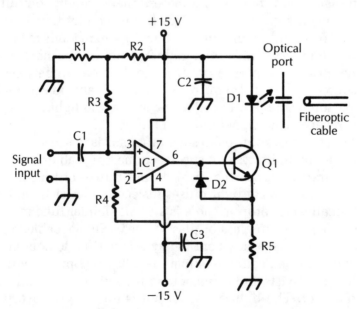

Fig. 12-22 *This remote transmitter circuit is designed for use with a fiberoptic cable.*

Table 12-13   Parts List for the Fiberoptic Transmitter of Fig. 12-22.

IC1	LF356 op amp
Q1	NPN transistor (2N2222 or similar)
D1	LED
D2	1N914
C1–C3	0.1-μF capacitor
R1	220-Ω resistor
R2	1.2-kΩ resistor
R3, R4	1-kΩ resistor
R5	47-Ω resistor

fiberoptic transmission. The parts list for this project is given in Table 12-13.

The modulated signal can be as high as 3.5 MHz (3,500,000 Hz). When no input signal is applied, the LED current will be 50 mA (0.05 A). The input signal should be kept within the 0- to +5-V range. The current across the LED will vary from 0 to 100 mA (0.1 A).

PROJECT 68:
Fiberoptic receiver

A fiberoptic receiver to use with the last project is shown in Fig. 12-23. The parts list is given in Table 12-14. The sensitivity of the

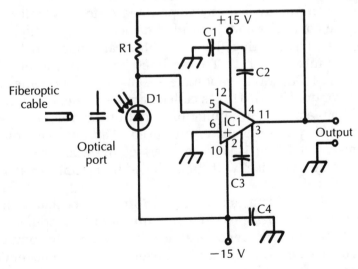

Fig. 12-23 *This is the receiver for the fiberoptic transmitter circuit of Fig. 12-22.*

Table 12-14   Parts List for the Fiberoptic Receiver of Fig. 12-23.

IC1	LH0032 op amp
D1	Photodiode (HP5082-4220)
C1, C4	0.1-μF capacitor
C2	1000-pF capacitor
C3	4.7-pF capacitor
R1	100-kΩ resistor

photodiode listed in the parts list is 0.5 A/W. The op amp boosts this small signal to a usable level which is passed on to the output. This system can be used for simple on/off control, or a modulator and demodulator can be added to allow multichannel control or voice communications.

Carrier-current control

It is often convenient not to have connecting wires between the controller and the controlled device in a remote control system. However such systems are rarely truly wireless. Except when portability is an absolute must, batteries tend to be too expensive, bulky, and generally inconvenient for use in a home control system. Most practical devices will have a power supply that derives its operating (dc) voltage from the ac electrical wiring that is run through the halls of almost all modern buildings.

You can't really call any ac-powered system "wireless" because the controller and controlled device are connected to the same (usually) ac wiring. There is a physical connection between the two units. So what? The ac lines just carry the operating power. True. But we can modulate the 60-Hz ac power line signal with a control signal. This is called carrier-current control.

There are certain limitations to this convenient system, which is why it isn't universally used. The designer must be extremely careful to isolate any part of the circuitry the user might come in contact with from the ac line. (Optoisolators are often utilized.)

Another problem stems from the simple fact that ac house wiring was never designed for communication of modulated signals. Noise and interference is almost inescapable. In some cases it can be so severe that the carrier-current system can be rendered useless. Such problems may or may not be temporary. It all

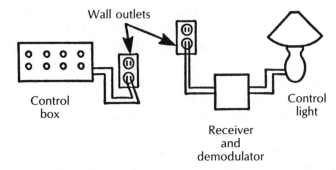

Fig. 12-24 *Your ac house wiring can be used to transmit control signals between remote devices.*

depends on what else happens to be using the same ac power lines.

Even so, the convenience of carrier-current control certainly makes it a viable alternative to consider when designing a control system. A typical carrier-current control system is illustrated in Fig. 12-24.

Now, let's consider how to put the idea of carrier-current control into actual practice. Figure 12-25 shows a simplified diagram of a home wiring system. Notice that there are actually two separate circuits here. The power line entering the home is a cable made up of three wires. One of these is a common line. The voltage between either of the outer wires with the common is approximately 120-V ac. (The actual voltage varies from time to time, depending on the current load on the system, and from location to location, depending on the specific equipment used by the power company.) Simple math tells us that the voltage between the two outer lines must be about 240-V ac.

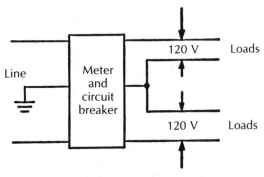

Fig. 12-25 *This is a simplified diagram of a typical home wiring system.*

You may run into problems with a carrier-current remote control system. The transmitter can be plugged into one line and the receiver can be plugged into the other. There might not always be a direct connection available. In large buildings with multiple fuse boxes, there might even be more than two circuits to contend with. A second potential problem is that there is no way to determine what else might be plugged into the circuit when the system is in use.

Many of these seemingly impossible problems can be dealt with by selecting the proper frequency for the carrier-current signal. When this fails, it is often possible to condition the wiring so that it will carry the control signals.

There is no way to specify a general purpose optimum frequency for a carrier-current system. Every wiring situation will be different. A few rules might be suggested, however.

A high-frequency (RF) transmission line should be terminated in its characteristic impedance, or potentially problematic standing waves can be created along the length of the line. Standing waves mean that the voltage and current will vary throughout the line. For lower-frequency signals, this isn't such a problem. The line length is not particularly important with low-frequency signals.

This is not to imply that a low-frequency system is problem free. At low frequencies the ac line looks like a random network of various resistances, capacitances, and inductances. There is no way to predict the exact electrical characteristics of the ac line. But these problems also exist at higher frequencies, along with the standing wave problem. In other words, using a low-frequency signal will usually be the lesser of two evils. If high frequencies are used, there might also be problems stemming from RF interference with broadcast signals.

Now, what low frequency should be used? We can narrow the choices down a little further. The ac power lines in the United States carry a strong (very strong) 60 Hz. This means you should not use a signal close to 60 Hz. It will just get lost in the power signal. By the same token, you should avoid the harmonics of 60 Hz (120 Hz, 180 Hz, 240 Hz, 300 Hz, 360 Hz, etc.). Any low frequencies in this range will have to be extremely powerful not to get buried in the noise.

The carrier-current signal should not be too low or too high in frequency. While the wiring situation can vary considerably, generally you'll have the best chance of success using frequencies in the 60- to 180-kHz range. This is high enough that the harmonics from the power line should be reduced to an insignificant level, but low enough that RF interference and transmission line problems shouldn't give you too much grief. For our discussion we will arbitrarily select a carrier frequency of 100 kHz, primarily because it's a nice neat value.

The carrier frequency is not the only frequency of importance in a carrier-current system. Control signals are tone encoded. The encoded tones are used to modulate the carrier signal. Either amplitude modulation (AM) or frequency modulation (FM) can be used. FM systems are more complex and expensive, but tend to be less noisy. (Tone encoding will be discussed in more detail later in this chapter.) The tone encoding frequencies should be selected to avoid 60 Hz and its harmonics, to limit interference from the power line.

If you need several control frequencies, you can save yourself a lot of trouble and calculating by adopting the dual-tone frequencies used in Touch Tone™ telephones. These frequencies were carefully chosen after a great deal of study. They are probably as close to optimum as you're likely to get.

In order to deal with any problems and spurious signals that might be in the power lines, you obviously need to know what they are. This is easier said than done in many cases. For one thing, making measurements is made difficult by the huge 60-Hz power signal. This will be tens or even hundreds of times stronger than potentially significant interfering signals.

If you try to measure the signal on the ac power line with an oscilloscope or other test instrument, all you'll see is a 60-Hz sine wave. Not very informative. You must filter out the 60 Hz signal (and possibly some of its lower harmonics) in order to accurately and meaningfully monitor the ac power line. You do this with a narrow band-reject (or notch) filter. The simplest type is illustrated in Fig. 12-26. This is called a twin-T network (or parallel T) because the schematic looks like it's made up of two Ts. One is comprised of C1, C2, and R1, while the second contains R2, R3, and C3. To center the rejection band around 60 Hz, the following component values can be used:

- R1 = 147 Ω
- R2, R3 = 295 Ω
- C1, C2 = 9 μF (at least 600 V)
- C3 = 18 μF (at least 600 V).

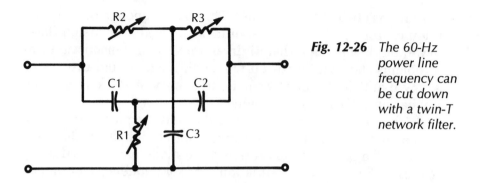

Fig. 12-26 *The 60-Hz power line frequency can be cut down with a twin-T network filter.*

The rather odd resistance values can best be obtained with trimpots (small, screwdriver-adjusted potentiometers).

The requirements for maximum attenuation at the center frequency are as follows:

- R1 = 147 Ω
- R2, R3 = 295 Ω
- C1, C2 = 9 μF (at least 600 V)
- C3 = 18 μF (at least 600 V)

The rather odd resistance values can best be obtained with trimpots (small, screwdriver-adjusted potentiometers).

The requirements for maximum attenuation at the center frequency are as follows:

R2 = R3 = 2 × R1
C1 = C2
C3 = 2 × C1

$$R1 = \frac{1}{2 \pi \, C3F}$$

$$C1 = \frac{1}{2 \pi \, R2F}$$

Where

F = desired center frequency and
Pi (π) = constant with an approximate value of 3.14.

Pi turns up quite frequently in electronic equations.

The trimpots must be tuned precisely because the rejection band is quite narrow and any deviation from the calculated values will alter the center frequency. You cannot attenuate the frequency you want to cut down.

PROJECT 69:
Carrier frequency generator

A carrier-current transmitter is generally made up of three primary stages, as shown in Fig. 12-27:

- carrier frequency generator
- transmitter
- coupling network.

A typical carrier frequency generator circuit is illustrated in Fig. 12-28. The parts list is given in Table 12-15. Basically this circuit is a VCO (voltage-controlled oscillator). The input signal modulates the output frequency. This is frequency modulation (FM).

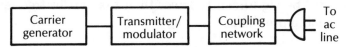

Fig. 12-27   *A carrier-current transmitter is made up of three basic stages.*

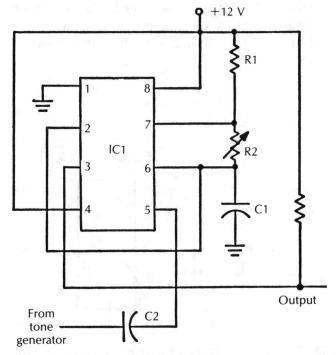

Fig. 12-28   *This is a typical carrier frequency generator circuit.*

Table 12-15   Parts List for the Carrier Frequency Generator of Fig. 12-28.

IC1	555 timer
C1	0.1-μF capacitor
C2	0.01-μF capacitor
R1	1-kΩ resistor
R2	100-kΩ potentiometer

The nominal (zero signal) output frequency can be calculated with the formula:

$$F = \frac{1.44}{(R1 + R2)C1}$$

The component values listed in Table 12-15 allow output frequencies from about 1 kHz to over 100 kHz. The input frequencies should be significantly lower than the carrier frequency to avoid aliasing problems.

Coupling network

The coupling network scarcely calls for a project of its own. This is just a twin-T filter, as discussed earlier in this chapter, and a few coupling capacitors. A simple coupling network circuit is illustrated in Fig. 12-29. Suitable component values are listed in Table 12-16.

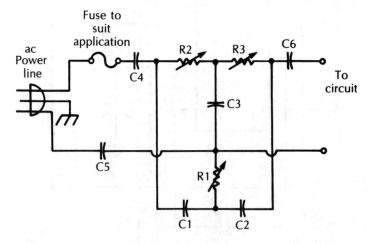

Fig. 12-29 *This simple coupling network can be used with the following carrier-current transmitter projects.*

Table 12-16 Parts List for the Coupling Network Circuit of Fig. 12-29.

C1, C2, C4, C5, C6	0.1-μF, 600-V capacitor
C3	0.2-μF, 600-V capacitor (may be two 0.1-μF capacitors in parallel)
R1	26,539-Ω resistor (50-kΩ potentiometer)
R2, R3	53,078-Ω resistor (100-kΩ potentiometer)

PROJECT 70:
First carrier-current transmitter

Now comes the transmitter section. One suitable circuit is shown in Fig. 12-30. The parts list for this project is given in Table 12-17.

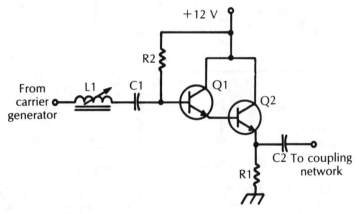

Fig. 12-30   *Carrier-current transmitter circuit 1.*

Table 12-17 Parts List for the Carrier-Current Transmitter of Fig. 12-30.

Q1, Q2	NPN transistor (2N4401 or similar)
L1	5- to 30-mH coil*
C1	0.0002-μF capacitor
C2	0.1-μF capacitor (500 V or more)
R1	1-kΩ resistor
R2	75-kΩ resistor

*The coil is adjusted to provide maximum drive for Q1. (A TV width coil may be used for L1.)

This circuit presents a very low impedance to the ac power line, which is highly desirable. The same basic circuit can be used with either AM or FM signals.

If the carrier-current signal must be transmitted over a long distance (say, in a large building), more powerful transistors than the ones specified in the parts list might be required.

PROJECT 71:
Second carrier-current transmitter

An alternate carrier-current transmitter circuit is shown in Fig. 12-31, with the parts list given in Table 12-18. This circuit is suitable for low-power signals.

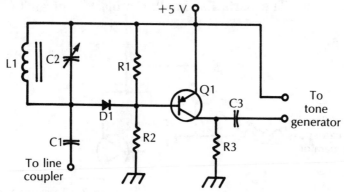

Fig. 12-31   Carrier-current transmitter circuit 2.

Table 12-18   Parts List for the Carrier-Current Transmitter of Fig. 12-31.

Q1	PNP transistor (GE53 or similar)
D1	SK3091 diode (or similar)
L1	Ferrite-rod antenna*
C1	0.1-μF, 600-V capacitor
C2	730-pF variable capacitor
C3	0.012-μF capacitor
R1	10-kΩ resistor
R2	220-kΩ resistor
R3	47-kΩ resistor

*Can be salvaged from an old AM radio.

PROJECT 72:
Third carrier-current transmitter

A final carrier-current transmitter circuit is illustrated in Fig. 12-32. It is built around the 555 timer. The complete parts list for this project is given in Table 12-19.

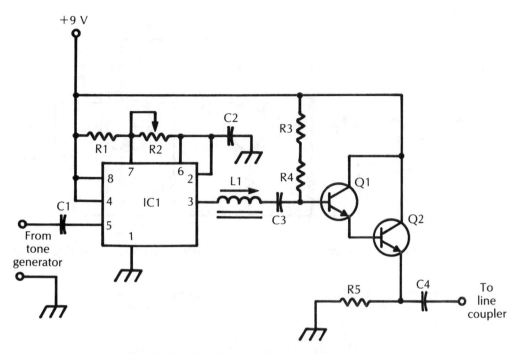

Fig. 12-32   *Carrier-current transmitter circuit 3.*

Table 12-19   Parts List for the Carrier-Current Transmitter of Fig. 12-32.

IC1	555 timer
Q1, Q2	NPN transistor (SK3122 or similar)
L1	32-mH adjustable coil
C1	0.01-μF capacitor
C2, C4	0.1-μF capacitor
C3	250-pF capacitor
R1	100-Ω resistor
R2	100-kΩ potentiometer (tune)
R3	68-kΩ resistor
R4	6.8-kΩ resistor
R5	1-kΩ resistor

PROJECT 73:
First carrier-current receiver

Obviously, a carrier-current transmitter isn't going to be good for very much beyond paperweight duty without a comparable receiver unit. Fig. 12-33 shows a receiver circuit intended for use

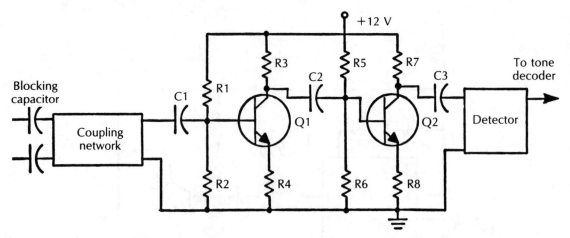

Fig. 12-33   *This carrier-current receiver is designed to work with the transmitter of Fig. 12-30.*

Table 12-20   Parts List for the Carrier-Current Receiver of Fig. 12-33.

Q1, Q2	NPN transistor (2N2222 or similar)
C1, C2, C3	0.01-μF capacitor
T1	10-kΩ:3.2-Ω impedance matching transformer
R1, R5	100-kΩ resistor
R2, R6	10-kΩ resistor
R3, R7	22-kΩ resistor
R4, R8	1.2-kΩ resistor

with the carrier-current transmitter of Fig. 12-30. It can be used with other transmitters as well.

Reception can be improved by adding an impedance matching network at the input. It would not be very practical to go into specifics on this here, because the requirements will vary from location to location. The tone decoder will be shown shortly. The parts list for this project is given in Table 12-20.

PROJECT 74:
Second carrier-current receiver

A receiver circuit suitable for the transmitter of Fig. 12-31 is shown in Fig. 12-34. The parts list for this project is given in Table 12-21.

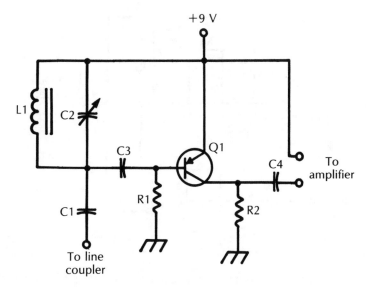

Fig. 12-34 *This circuit is another carrier-current receiver.*

Table 12-21   Parts List for the Carrier-Current Receiver of Fig. 12-34.

Q1	PNP transistor (GE-53 or similar)
L1	Ferrite-rod antenna*
C1	0.1-μF, 600-V capacitor
C2	730-pF variable capacitor
C3	0.18-μF capacitor
C4	0.12-μF capacitor
R1	100-kΩ resistor
R2	1-kΩ resistor

*Can be salvaged from an old AM radio.

PROJECT 75:
Tone decoder

A simple tone decoder circuit is illustrated in Fig. 12-35. It may be used with either of the receiver circuits presented so far (Figs. 12-33 and 12-34).

As the parts list in Table 12-22 indicates, this circuit is built around the 567 tone decoder IC, which is manufactured for just this type of application. More information on tone decoders is given later in this chapter.

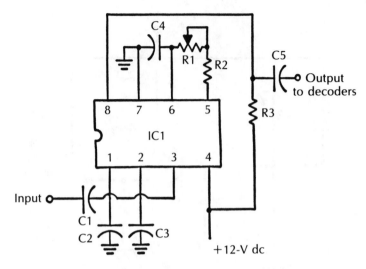

Fig. 12-35 *This simple tone decoder can be used with either of the carrier-current receiver circuits (Figs. 12-33 and 12-34).*

Table 12-22   Parts List for the Tone Decoder of Fig. 12-35.

IC1	567 tone decoder IC
C1	0.1-μF capacitor (500 V or more)
C2, C5	0.1-μF capacitor
C3	25-μF, 35-V electrolytic capacitor
C4	0.005-μF capacitor
R1	20-kΩ potentiometer
R2	2.2-kΩ resistor
R3	4.7-kΩ resistor

PROJECT 76:
Third carrier-current receiver/tone decoder

One final carrier-current receiver circuit is illustrated in Fig. 12-36. This circuit incorporates its own built-in tone decoder. The parts list for this project is given in Table 12-23.

Tone encoding

Using different frequencies for different functions is a convenient way to combine multiple control signals on a single channel. Almost any transmission method can be used, including direct connection through wires, RF signals, optical signals, or ac car-

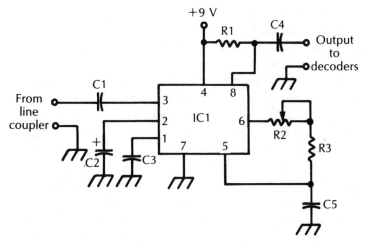

Fig. 12-36 *This circuit combines a carrier-current receiver and a tone decoder.*

**Table 12-23 Parts List for the
Carrier-Current Receiver/Tone Decoder of Fig. 12-36.**

IC1	567 tone decoder IC
C1	0.1-μF, 600-V capacitor
C2	22-μF, 35-V electrolytic capacitor
C3, C4	0.1-μF capacitor
C5	0.005-μF capacitor
R1	3.9-kΩ resistor
R2	25-kΩ potentiometer
R3	22-kΩ resistor

rier currents. Another transmission method is sonic. Just send out the sounds with some kind of loudspeaker and pick them up with some kind of microphone.

In some cases, audio control tones can be annoying or distracting. Ultrasonic signals (above 20 kHz) will function in pretty much the same way, but are not audible to humans. (They might drive your dog or cat crazy though.) Ultrasonic transducers (speakers and microphones) might be more expensive and harder to find than AF (audio frequency) devices.

Very simple on/off control can be accomplished with a simple burst of sound and a VOX (voice-operated switch) circuit. However, it might succumb to false triggering because of environmental noises.

Better accuracy can be achieved by assigning a specific frequency to the control function. The controlled device is designed to recognize only its assigned frequency. Sounds at other frequencies are ignored.

Once you've gone this far, why not assign a second frequency to a second control function? Then add a third, a fourth . . . the possibilities are virtually limitless. This is called tone encoding. The tones are generated by a circuit called a tone encoder. The receiver has a tone decoder circuit to decide what signal will do what. A simple tone-operated system is illustrated in block diagram form in Fig. 12-37. The output of each tone decoder is usually a simple dc control signal.

Tone encoding is a simple matter. Just use one or more oscillator circuits and gate the output on and off as desired to activate the specific control function.

Tone decoding is a little more complex. There are two basic approaches to tone decoding—band-pass filters and PLLs (phase-locked loops). Each of these approaches will be discussed in the following pages.

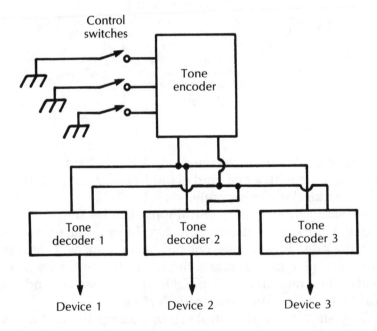

Fig. 12-37 *A simple tone-operated system permits multiple functions to be controlled from a single channel.*

Filter decoding

Probably the most obvious approach to tone decoding is the filter method. A band-pass filter is a circuit that passes only those frequencies which lie within a specific band (or range). Any signals with frequencies outside that band will not reach the output. A graph illustrating the action of an ideal band-pass filter is shown in Fig. 12-38.

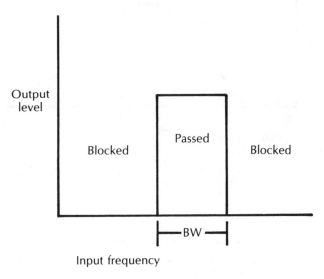

Fig. 12-38 *An ideal band-pass filter rejects all other frequency signals that are outside its specified pass band.*

Practical circuits cannot distinguish quite so sharply between frequencies. Instead of a sharp, instantaneous cutoff, there will be a more gradual slope. The frequency response of a practical band-pass filter is illustrated in Fig. 12-39.

The action of a band-pass filter is defined primarily by two specifications. The center frequency (F_c) is the midpoint of the passed band. The bandwidth (BW) is the size of the passed band; that is, how many frequencies are passed.

For example, let's consider an ideal band-pass filter that passes only those frequencies between 2300 and 3700 Hz. The center frequency is 3050 Hz, and the bandwidth is 3700 − 2300 = 1400 Hz.

Frequencies close to, but outside, the pass band of a practical filter will get through to the output, but will be attenuated to a degree proportional to the distance from the pass band.

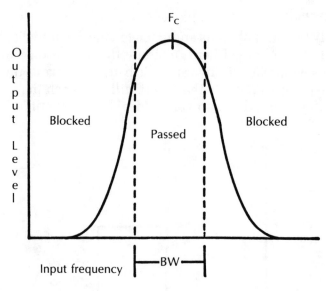

Fig. 12-39   *A practical band-pass filter has a gradual slope between the pass band and the rejection bands.*

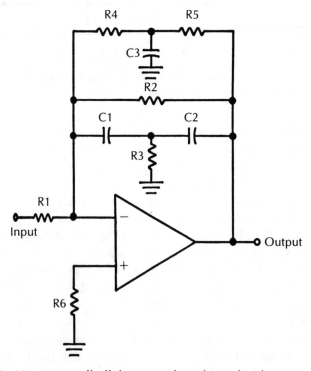

Fig. 12-40   *Very steep roll-off slopes can be achieved with an op amp based filter circuit.*

Obviously, the steeper the roll-off slope, the better the selectivity of the filter will be. Very steep roll-off slopes are possible with op amp based circuitry, such as the one illustrated in Fig. 12-40. The bandwidth of this circuit is quite narrow. The center frequency (F_c) is determined by the values of coil L and capacitor C, according to the formula

$$F_c = \frac{1}{6.28 \times \sqrt{L \times C}}$$

Let's assume that L = 1500 μH (0.0015 H), and C = 0.0022 μF (0.0000000022 F). In this case, the center frequency works out to

$$F_c = \frac{1}{6.28 \times \sqrt{0.0015 \times 0.0000000022}}$$
$$= \frac{1}{6.28 \times \sqrt{0.0000000000033}}$$
$$= \frac{1}{6.28 \times 0.0000018}$$
$$= \frac{1}{0.0000114}$$
$$= 87{,}656 \text{ Hz}$$
$$\approx 88 \text{ kHz}$$

In practical design applications, you will usually know the desired center frequency, and you'll need to determine the component values. Because coils are available in a lesser range of common values, it makes sense to start out by arbitrarily selecting a likely inductance value. Then we rearrange the equation to solve for the capacitance:

$$C = \frac{1}{L} \times \left(\frac{1}{6.28 \times F_c} \right)^2$$

You will often have to try several combinations to find practical values for both components. As an example, let's say we need a band-pass filter with a center frequency of 50 kHz (50,000 Hz). Let's try a 1000-μH (0.001 H) coil for L. What value capacitor should we use?

$$C = \frac{1}{0.001} \times \left(\frac{1}{6.28 \times 50{,}000} \right)^2$$
$$= 1000 \times \left(\frac{1}{314000} \right)^2$$
$$= 1000 \times (0.0000032)^2$$
$$= 1000 \times 0.00000000001$$
$$= 0.00000001 \; F$$
$$= 0.01 \; \mu F$$

There are other components in the circuit, besides the capacitor and coil. The resistor values determine the gain of the passed frequencies. The formula is

$$G = -\frac{R2}{R1}$$

The negative sign simply indicates that the output will be 180 degrees out of phase with the input. In most control applications this will be totally irrelevant.

Once again, we'll make things clearer with a simple example. Let's say that R1 = 22 kΩ (22,000 Ω), and R2 = 220 kΩ (220,000 Ω). The gain would be equal to

$$G = -\frac{220{,}000}{22{,}000}$$
$$= -10$$

An input of 1 mW should result in an output of 10 mW, as long as its frequency is within the pass-band. Resistor R3 should have a value equal to or slightly less than that of R1.

Other band-pass filter circuits can also be used, but we don't really need to go into them here. Filter design is covered extensively in many other electronics texts.

In a multichannel filter decoding receiver, several separate band-pass filters are used, one for each encoded frequency (and for each independent controlled function). A block diagram of such a system is illustrated in Fig. 12-41. The control signal triggers the appropriate device if (and only if) there is an output from the appropriate filter. Obviously, this happens only when the input frequency is within that filter's pass band.

Each encoded frequency must be carefully selected so that it is within the pass band of one and only one of the filters. Remem-

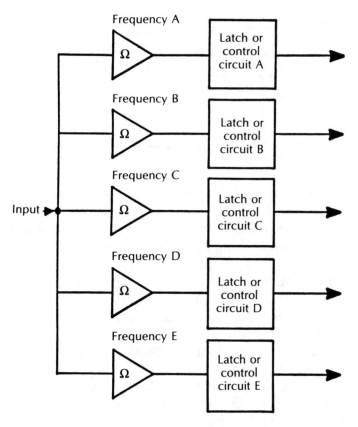

Fig. 12-41 *In a multichannel filter decoding receiver, several separate band-pass filters are used; one for each encoded frequency.*

ber the gradual nature of the roll-off slope. Even a partially attenuated signal within the slope region can cause false triggering. Be very, very careful when choosing your encoding frequencies. If the proper encoding frequencies are selected, such a system can be reasonably reliable and efficient. Only a relatively small number of control lines will be practical for most filter decoding systems.

PLL decoding

The other method of tone control uses the PLL (phase-locked loop). PLLs have a largely undeserved reputation for being extremely complicated. Actually, they are conceptually quite similar to the closed-loop automation system discussed at several points throughout this book. The basic components of a PLL are illustrated in block diagram form in Fig. 12-42. As in the closed-

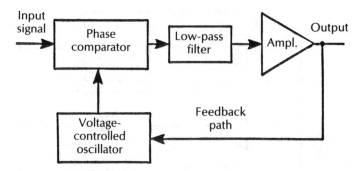

Fig. 12-42   *A PLL is a simple closed-loop system.*

loop automation system, the key to the PLL is the feedback loop.

A PLL circuit is made up of three basic stages:

- phase comparator
- low-pass filter/amplifier
- VCO (voltage-controlled oscillator).

In most modern cases, all three stages are generally contained within a single IC chip.

Two signals are fed into the phase comparator. One is the external input signal, the other is the output of the VCO (and the PLL itself)—this is the feedback loop.

If these two signals are perfectly in step with each other, and 90 degrees out of phase, there will be no output from the phase comparator. The output of the PLL (the VCO's frequency) will not be changed.

If the two signals are not locked together, however, the phase comparator will produce an output called the error signal. This signal is fed through a low-pass filter for smoothing and to prevent oscillation within the closed-loop system. The error signal is then amplified and fed to the control input of the VCO, changing its phase and possibly its frequency, until it is locked onto (and 90 degrees out of phase with) the external input signal.

A popular PLL IC is the 567. It was designed with frequency sensing and tone decoding applications in mind. A block diagram of this device is shown in Fig. 12-43. Notice that the 567 also contains a second (quadrature) phase detector which is fed to a power output stage. When the input frequency is close to the center frequency of the device (determined by external compo-

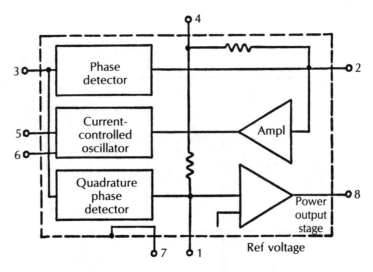

Pin connections

1. Output filter capacitor   5. Timing resistor
2. Low-pass filter capacitor   6. Timing resistor and capacitor
3. Input   7. Ground
4. Supply voltage   8. Output

Fig. 12-43 *A popular PLL device for tone decoding applications is the 567 IC.*

nents), pin 8 goes to ground level (0 V). At all other times, this pin will be kept floating.

A simple tone decoder circuit built around the 567 is illustrated in Fig. 12-44. Resistor R1 and capacitor C1 set the free-running (center) frequency of the current-controlled oscillator. With the component values shown, this frequency will be about 50 kHz (depending on the setting of the potentiometer). This will be the frequency detected by the tone detector. Unless the external input frequency is close to the oscillator frequency, pin 8 will be floating and no current will flow through the load. If the correct frequency is applied to the input, the quadrature phase detector turns on the power output stage, grounding pin 8, and allowing a current (of up to 100 mA) to flow through the load.

Noise immunity of this tone decoder is achieved by adjusting the amount of time required for it to respond to a tone. Slower reaction times allow the circuit to ignore brief transients that might otherwise confuse it. The response time can be adjusted by changing the circuit's bandwidth. The wider the bandwidth, the quicker the response. For slower, more noise-immune response, a

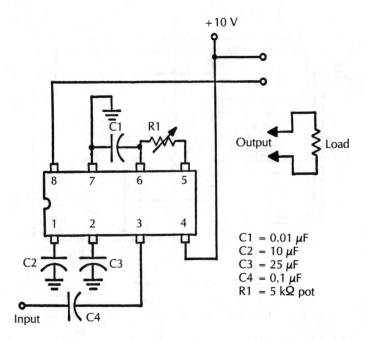

Fig. 12-44 *This is a typical tone decoder circuit built around the 567 PLL IC.*

narrow bandwidth should be used. The bandwidth is determined by several factors, primarily, the applied signal voltage, the center frequency, and the value of capacitor C2.

The center frequency, as mentioned earlier, is set by the values of R1 and C1. The formula is

$$R1 = \frac{1}{F_c C1}$$

The usable range of center frequencies for the 567 runs from 0.01 Hz to 500 kHz (500,000 Hz). This range should take care of almost any control application you're likely to need.

Touch-tone encoding

If you are using more than just a few control functions in a tone encoding system, greater reliability can be achieved with complex tones. Each frequency must be selected to avoid harmonics that can cause false triggering and to avoid 60 Hz (the power-line frequency) and its harmonics.

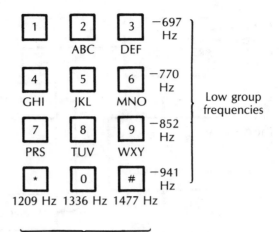

Fig. 12-45 *The Touch Tone ™ system is made up of 12 control signals arranged in a column/row matrix.*

The phone company has already done a lot of work in this area, so you might as well take advantage of their extensive research and use the Touch Tone™ system. Each digit is represented by a combination of two tones. Both tones must be present for a valid control signal to exist. This greatly limits the chances of false triggering.

A big advantage of using the Touch Tone™ system is the availability of preconstructed devices. A standard Touch Tone™ telephone set can be used to generate all of the tones. Circuits and ICs for generating the Touch Tone™ signals are also widely available and can be easily incorporated into your control projects.

The Touch Tone™ system is made up of 12 control signals arranged in a column/row matrix, as illustrated in Fig. 12-45. Whenever one of the buttons is depressed, a low frequency and a high frequency are simultaneously generated.

Several manufacturers make ICs for use in Touch Tone™ systems. A typical example is the MC14410 tone encoder IC from Motorola. This device is illustrated in Fig. 12-46. All of the component tones are derived from a 1-MHz crystal oscillator. This base frequency is divided digitally to create each of the component frequencies. An additional column is supported by this device, permitting a total of 16 control switches, as shown in Fig. 12-47. A tone-encoding circuit built around the MC14410 is illustrated in Fig. 12-48.

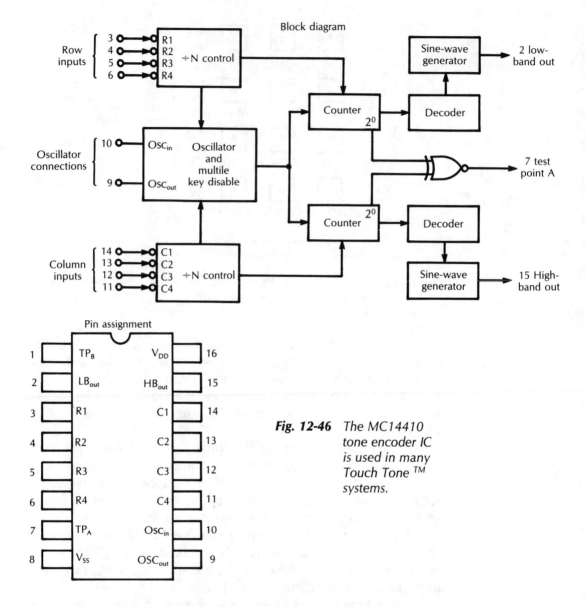

Block diagram

Pin assignment

Fig. 12-46 *The MC14410 tone encoder IC is used in many Touch Tone ™ systems.*

At the receiving end, a tone decoder for each of the seven (or eight) component frequencies must be provided. Gates are used to combine the tone decoder outputs appropriately. A circuit for a standard 4 × 3 keypad system (12 signals, 7 frequencies) is shown in Fig. 12-49. The output of each NOR gate will be high if and only if both of its inputs are low. Each input goes low when the appropriate tone decoder is turned on (pin 8 of the 567 grounded).

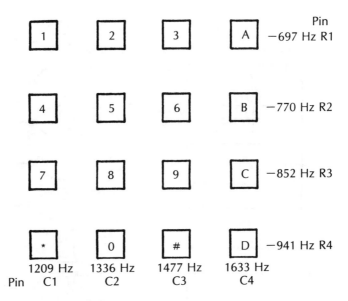

Fig. 12-47 *An additional column can be added to the Touch Tone ᵀᴹ system for a total of 16 control signals.*

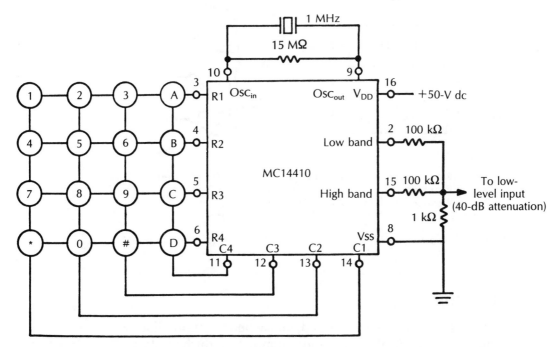

Fig. 12-48 *This is a typical Touch Tone ᵀᴹ encoding circuit built around the MC14410.*

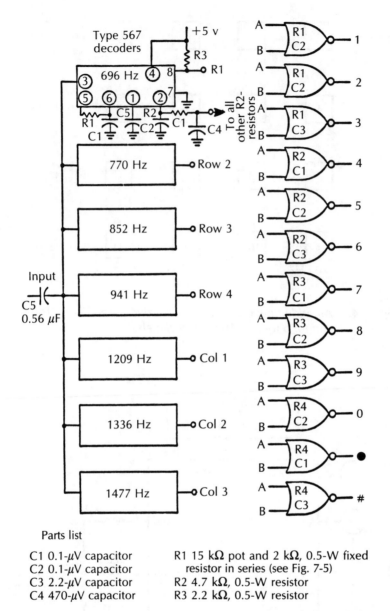

Parts list

C1 0.1-μV capacitor	R1 15 kΩ pot and 2 kΩ, 0.5-W fixed
C2 0.1-μV capacitor	resistor in series (see Fig. 7-5)
C3 2.2-μV capacitor	R2 4.7 kΩ, 0.5-W resistor
C4 470-μV capacitor	R3 2.2 kΩ, 0.5-W resistor

Fig. 12-49   *This is a receiver/decoder circuit for the tone encoder circuit of Fig. 12-48.*

Radio control

When wireless control is mentioned, most people immediately think about radio control. While radio control might be a desirable choice in some applications, in most cases it introduces more trouble than it's worth. Interference problems are almost inevitable. Radio circuits are usually quite critical in terms of component selection, placement, and shielding. Radio frequency circuitry tends to be rather expensive. In addition, you can easily (perhaps inevitably) run into problems with legal restrictions on radio transmissions. Still, despite these problems, there are certainly applications in which radio control may be the best (occasionally the only practical) choice, so we can't afford to ignore it here.

The first things to consider are the legal restrictions. All RF (radio frequency) transmitters come under the jurisdiction of the FCC (Federal Communications Commission). Don't try to cheat here. The hefty fines if you get caught make this a very poor gamble. The FCC's detection equipment is surprisingly accurate, and many an experimenter has been caught violating the FCC's rules and regulations.

Don't suffer from the misapprehension that if the power level is low enough it is not considered a radio transmitter. If it transmits RF signals at all, it is a radio transmitter. The power level is irrelevant to the definition of a radio transmitter.

In most cases, radio transmitters must be licensed by the FCC to be legally used. There are a few exceptions, which are described in part 15 of the FCC rules. (If you are considering using radio control, you should carefully read the FCC rules, which are available from the local branch office of the FCC.)

The FCC assigns specific functions to various frequency bands in the radio spectrum. Three bands have been assigned to radio control devices:

- 27 MHz
- 50 to 54 MHz
- 72 to 76 MHz.

The highest (72 to 76 MHz) and the lowest (27 MHz) of these bands do not require an operator's license. This is not to imply that there aren't any legal restrictions on their use. Read the FCC rules before using these frequency bands for a control system.

Each band is divided into several channels:

- 27 MHz band
  —26.995 MHz
  —27.045 MHz

—27.095 MHz
—27.145 MHz
—27.195 MHz
—27.225 MHz
• 72 to 76 MHz band
—72.08 MHz*
—72.16 MHz
—72.24 MHz*
—72.32 MHz
—72.40 MHz*
—72.96 MHz
—75.64 MHz*

The four frequencies in the 72- to 76-MHz band that are marked with asterisks (*) are restricted to use with model aircraft only. Any other use is illegal.

An amateur radio technician class license (or better) is required for the 50 to 54 MHz band. The assigned channels in this band are

• 51.20 MHz
• 52.04 MHz
• 53.10 MHz
• 53.20 MHz
• 53.30 MHz
• 53.40 MHz
• 53.50 MHz.

Notice that many of the channel assignments in each band are closely spaced. This means that highly selective receivers (usually superheterodyne type) must be used.

Probably the easiest approach to using radio control in your projects is to use a transmitter from some commercial device such as a garage door opener or model airplane. These transmitters are already type approved, type accepted, or certified by the FCC. While you can modify the receiver any way you like to suit your application, it is illegal to make any modification to such a transmitter. That includes adding a longer antenna! If you are unsure about the legality of anything you might have in mind, check with the local branch office of the FCC, or change your plans. Do no risk a large fine or possible jail sentence.

If possible, you should avoid opening the transmitter housing unless you are licensed to work on radio transmitters and definitely know what you're doing. Actually, this isn't a significant limitation for most control applications. A control signal is a control signal. It's what the receiver does with the control signals that matters. There are no legal restrictions on modifying a receiver.

PROJECT 77:
Low-power AM transmitter

It is legal to build a small unlicensed transmitter for the AM broadcast band (54 to 160 kHz). The transmitted signal cannot cover more than a 50-ft. range.

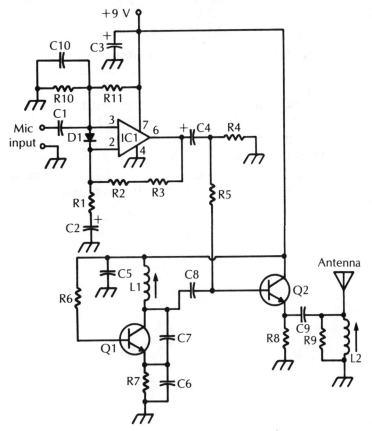

Fig. 12-50 *This small transmitter will send signals to any AM radio.*

**Table 12-24   Parts List for the
Low-Power AM Transmitter Circuit of Fig. 12-50.**

IC1	CA3140 op amp
Q1, Q2	NPN transistor (2N3904 or similar)
D1	1N914 diode
C1	0.22-μF capacitor
C2	1-μF, 35-V electrolytic capacitor
C3	5-μF, 35-V electrolytic capacitor
C4	10-μF, 35-V electrolytic capacitor
C5	0.05-μF capacitor
C6	1000-pF capacitor
C7	500-pF capacitor
C8	33-pF capacitor
C9	270-pF capacitor
C10	330-pF capacitor
R1	4.7-kΩ resistor
R2, R3	1-MΩ resistor
R4, R5, R6	2.2-kΩ resistor
R7	560-Ω resistor
R8	270-Ω resistor
R9	33-kΩ resistor
R10, R11	100-kΩ resistor
L1, L2	Coils (selected for frequency)

A simple AM wireless microphone circuit is shown in Fig. 12-50. The parts list is given in Table 12-24. Instead of a microphone (voice) input, you can use control tones. The signal can be picked up by a nearby standard AM radio that is tuned to the transmitter frequency. Be aware that there might be considerable interference within this band, limiting its usefulness for control applications.

❖ 13
Computer control

THE ULTIMATE IN AUTOMATION SYSTEMS EMPLOYS COMPUTER control. A computer can be programmed for a wide variety of automated control patterns: simple or complex, open loop or closed loop. The biggest advantage of computer control lies in the programmability. The control pattern can be altered at any time, just by reprogramming the computer. No physical rewiring of the hardware circuitry is required.

Just a few short years ago, few hobbyists could reasonably afford to dedicate a computer to automation tasks. Now, microcomputers are widely marketed almost everywhere. Many are available for under $100. You can almost get one for each automation system, and not use it for anything else, and still get your money's worth.

Even if you already own a fancy microcomputer, you might want to consider buying a budget model for your automation systems. Automation applications will rarely call for heavy-duty computing ability, super-fast calculations, or large memory banks. A budget microcomputer will do the job just fine, and you won't have to tie up your main computer for the automation tasks.

If you have a strong background in electronics design, you might want to consider building a dedicated computer system. A CPU (central processing unit) can be purchased for $10 to $20. You don't need to pay for a fancy case or any circuitry you don't really need in your system.

If your technical expertise isn't quite up to the challenge of designing a computer-based system from scratch (or you just don't want to bother), you can accomplish pretty much the same things by using the I/O (input/output) ports of a full microcomputer. In this book we will concentrate primarily on this second, somewhat simpler approach. Much of what is said here will also apply (with minor modifications) to a complete home-brew CPU-based system.

Programming

Virtually every microcomputer in existence requires somewhat different programming than every other microcomputer. There will be some degree of overlap in techniques. We will just discuss the general basics in this chapter.

There are many different programming languages around today. These are analogous to languages like English or French or Turkish. A programming language is just the way you tell the computer what you want it to do.

Ultimately, any computer only understands one language. This is called machine language. (Different CPUs usually understand different forms of machine language. They can't directly "talk" to each other.) Machine language is made up of binary numbers. Binary numbers contain only 1s and 0s. You can think of each binary word or "byte" as a string of switches, as illustrated in Fig. 13-1. If a given switch is open, that binary digit (or

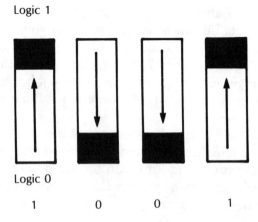

Logic 1

Logic 0

1   0   0   1

Fig. 13-1 *Binary words (bytes) can be thought of as a series of switches.*

"bit") is a 0. If the switch is closed, the bit is a 1. Different combinations of 1s and 0s mean different things to the CPU.

Fortunately, almost all commercial microcomputers come supplied with a built-in translator program, so the computer can understand a "higher" (more English-like) language. BASIC is by far the most popular language. This is not the place to go into the debate of whether or not it's the best choice. In today's market, it is definitely the most readily available and convenient choice. In some cases, it is the only available choice.

BASIC is so popular because it is such an easy language to learn and work with. English-like commands are used. In many cases, the meaning of the command is perfectly obvious. For example

PRINT 2 + 3

tells the computer to display a "5" on its video screen.

It would not be appropriate to cover BASIC (or any other programming language) in depth here. Many fine books on learning BASIC have been written.

Input signals

In most automation systems, the computer will need to sense various conditions in the outside world. Many sensors can be connected directly to an input port on the computer. Almost any device that performs a switching function can be used in this manner. The computer can tell if each switch connected to its input ports is open or closed.

Several switching-type sensors can be connected to a single input port. Each switching device is one bit of the incoming binary word. For most inexpensive microcomputers, each port can handle one byte (eight bits). Therefore, eight switching sensors can be fed into a single input port, and the computer will be able to individually distinguish between them.

To see how this is done, let's look at a typical example. The following devices are connected to input port A:

- bit 0—door intrusion switch (front)
- bit 1—window intrusion switch 1
- bit 2—window intrusion switch 2
- bit 3—window intrusion switch 3
- bit 4—door intrusion switch (back)

- bit 5—smoke detector
- bit 6—temperature sensor
- bit 7—flooding sensor (basement).

Now, the computer is programmed to periodically check the value at input port A. If it finds a value of 0000 0000, it assumes everything is OK (all switches are open) and moves on to other programming. Any other value at input port A indicates an alarm condition (one or more switches is closed). The computer is programmed to take an appropriate action depending on the specific value detected. For instance if the value at input port A is

1000 0000

the computer might display a message reading, "Flooding in Basement." In addition, it might feed out some control signals to the output ports (which work in the same, but opposite manner as the input ports). These control signals could activate a sump pump and possibly trigger an audible alarm.

Now let's say the incoming value at input port A is

0010 0000

This code would trigger an intrusion alarm. The computer can even pinpoint the location of the intrusion. For example, the display might read "Intrusion—Master Bedroom Window." A light could be turned on at the appropriate location to startle the intruder.

With a little more advanced programming and hardware, the computer could automatically dial the phone number of the local police department or a neighbor, then activate a tape recorder to play a prerecorded message giving your address and requesting assistance. As you can see, computer control can range from very simple to quite complex.

It is also possible for the computer to be programmed to respond only when a specific combination of input switch sensors are activated. For example, an input value of

0110 0001

might trigger a special action that is not activated by any other input value. The possibilities are limited only by your imagination and programming skill.

D/A conversion

A computer can recognize and put out only digital signals, for simple on-off functions. This is quite sufficient for a great many applications. Other real-world applications aren't quite so cooperative. A continuous range of values might be needed either as an input or an output signal.

A computer cannot input or output a continuous range analog signal. But all is not lost. It is certainly possible to convert between analog and digital signals. We will look at digital-to-analog (D/A) conversion first, because it is somewhat simpler.

Remember the individual bits of a byte do not have to be treated like independent entities. Each bit has a value dependent on its position within the byte. Each bit's value is a power of two. That is,

Bit	Value
0	1
1	2
2	4
3	8
4	16
5	32
6	64
7	128

Bit 0 is the right-most bit. Bit 7 is the left-most. That is,

$$7\text{-}6\text{-}5\text{-}4\text{-}3\text{-}2\text{-}1\text{-}0$$

Bit 7 is worth much more (has greater weight) than any of the other bits. Bit 0 is worth the least (the least weight).

To find the total value of a binary word, you just add together the place values of all bits that are 1s and ignore the 0s. For example

$$0101\ 1001 = 0 + 64 + 0 + 16 + 8 + 0 + 0 + 1 = 89$$

In other words, 0101 1001 is 89 times as great as 0000 0001.

We can try summing together the bits of an output port. Let's say each bit switches between 0 and + 1 V. In this case 0000 0001 would become an analog output of + 1 V, which is fine. But 0101 1001 would be a mere + 4 V, because all of the bits have equal

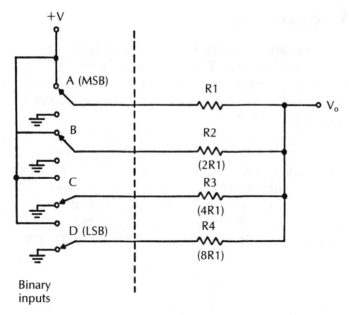

Fig. 13-2 *A simple D/A converter can be made with weighting resistances.*

weight. This is obviously not what we're looking for. We need some way to weight each bit.

A simple solution is to put a resistor with an appropriate value in series with each bit, as shown in Fig. 13-2. Only four bits are shown here for simplicity. The digital bits are represented as SPDT switches that individually select +V or 0 V.

The most significant bit (MSB—largest value) passes through the smallest resistance. Lesser bits pass through greater resistances that are multiples of two of the minimum (MSB) resistance value. Obviously, the least significant bit (LSB—smallest value) should pass through the largest resistance. This technique works in accordance with Ohm's law:

$$E = IR$$

If we choose a value of 10 kΩ for resistor R1, then R2 will be 20 kΩ, R3 will be 40 kΩ, and R4 will be 80 kΩ. Holding the current constant, the amount of voltage drop across each resistor will be weighted proportional to its binary value.

For our discussion, we will assume a logic 0 equals 0 V, and a logic 1 equals +5.0 V. We will assume the current drawn by the output load circuit gives the following voltage drops across each of the resistors (for a +5.0-V input):

- R1 = 1.0 V
- R2 = 3.0 V
- R3 = 4.0 V
- R4 = 4.5 V.

If all of the input bits are at logic 0, the output will obviously have to be 0 V. If one and only one of the input bits is at logic 1, the output voltage is simply the difference between the input voltage (+ 5.0 V) and the appropriate resistor. For instance, let's say the full input value is 0010. Only input C is a logic 1. Therefore, the output voltage will be

$$V_o = 5.0 - 4.0 = 1.0 \text{ V}$$

Each weighted bit works out to a voltage that is exactly twice its predecessor:

Bit	Voltage
D	0.5 V
C	1.0 V
B	2.0 V
A	4.0 V

This is the same weighting used in the binary numbering system.

Thanks to the consistency of the weighting scheme, intermediate values (two or more bits at logic 1) will also work out right:

Binary input	Decimal value	Output voltage		
0000	0	0 + 0 + 0 + 0	=	0 V
0001	1	0 + 0 + 0 + 0.5	=	0.5 V
0010	2	0 + 0 + 1 + 0	=	1.0 V
0011	3	0 + 0 + 1 + 0.5	=	1.5 V
0100	4	0 + 2 + 0 + 0	=	2.0 V
0101	5	0 + 2 + 0 + 0.5	=	2.5 V
0110	6	0 + 2 + 1 + 0	=	3.0
0111	7	0 + 2 + 1 + 0.5	=	3.5 V
1000	8	4 + 0 + 0 + 0	=	4.0 V
1001	9	4 + 0 + 0 + 0.5	=	4.5 V
1010	10	4 + 0 + 1 + 0	=	5.0 V
1011	11	4 + 0 + 1 + 0.5	=	5.5 V
1100	12	4 + 2 + 0 + 0	=	6.0 V
1101	13	4 + 2 + 0 + 0.5	=	6.5 V
1110	14	4 + 2 + 1 + 0	=	7.0 V
1111	15	4 + 2 + 1 + 0.5	=	7.5 V

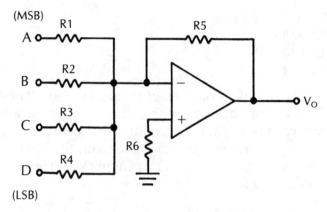

Fig. 13-3 *A practical direct weighting D/A converter includes a buffer-amplifier stage to limit loading problems.*

The output increases in 0.5-V (LSB) steps. The analog output is directly proportional to the digital input value. That's exactly the result we were looking for.

But there are still problems. Uneven loading by the output can cause the current draw to vary, which affects the voltage drops. To head off loading problems, a buffer amplifier stage should be added to the simple D/A converter, as illustrated in Fig. 13-3. This buffer stage can also amplify the signal, if desired.

This system is simple and direct, but it can have a number of problems if more than just a few input bits are to be converted. With more than just a few bits, the resistances will have to cover a fairly large range. Rather odd (not standardly available) values will be called for. In addition, the resistors have to be closely matched. Wide tolerances in the value can completely upset the accurate weighting of the bits.

The most obvious way to obtain the odd resistances required would be to put several standard-value resistances in series (or sometimes in parallel). Unfortunately, the tolerances add. Let's say we want to make a 40-kΩ resistance from four 10-kΩ resistors in series. If each 10-kΩ resistor has a tolerance of 5%, the total resistance will have a tolerance of 20%. That is, its value could be anything from 32 to 48 kΩ.

The wide resistance range can also be problematic. Consider the resistances for an eight-bit D/A converter of this type. If we use a 10-kΩ resistor (R) for the most significant bit (MSB), the other resistances will be as follows:

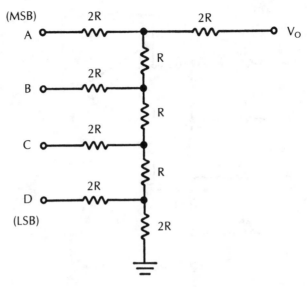

Fig. 13-4 *A better approach to D/A conversion is to use an R-2R ladder network.*

Input bit			Resistance value
7 (MSB)	R1	1R	10 kΩ
6	R2	2R	20 kΩ
5	R3	4R	40 kΩ
4	R4	8R	80 kΩ
3	R5	16R	160 kΩ
2	R6	32R	320 kΩ
1	R7	64R	640 kΩ
0 (LSB)	R8	128R	1280 kΩ = 1.28 MΩ

Obviously, this direct approach is awkward at best. A more convenient type of D/A converter is the R-2R ladder network that is illustrated in Fig. 13-4. Only two resistance values are needed—R and 2R. These two values are used repeatedly throughout the circuit. By using the combination scheme shown in the diagram, each incoming bit will be properly weighted. This system can be extended to accept as many bits as you need.

Practical R-2R ladder D/A converter circuits should also include a buffer amplifier stage to prevent loading problems. This is shown in Fig. 13-5.

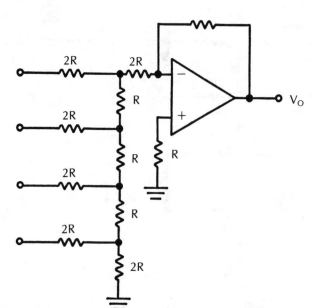

Fig. 13-5 *Practical R-2R D/A converters also include a buffer-amplifier stage.*

A/D conversion

In the last section, we adapted the computer's digital output to the analog world. In many practical applications (especially closed-loop automation systems) it is necessary for the computer to examine and interpret analog input data. The analog input signal must be converted to digital form to be usable by the computer.

This is done with an A/D (analog-to-digital) converter, which is just the opposite of the D/A converters described in the previous section. Unfortunately, A/D conversion is more complicated than D/A conversion. There are several possible approaches.

In most A/D converters, the input signal is sampled at a regular rate (typically several hundred or thousand times per second). Each sample is then individually converted into a proportional digital value.

The conversion isn't perfect, of course. Digital values change in a stepwise manner, rather than a smooth linear spectrum. Intermediate values must be rounded off. For example, if we set up a system of 1 V/bit, the recognized values are as follows:

0000	0 V
0001	1 V
0010	2 V
0011	3 V
0100	4 V
0101	5 V
0110	6 V
0111	7 V
1000	8 V
1001	9 V
1010	10 V
1011	11 V
1100	12 V
1101	13 V
1110	14 V
1111	15 V

If the sampled value is 3.5 V, the computer will have to see it as either 0011 (3.0 V) or 0100 (4.0 V). We are locked into the step units of the conversion factor.

If we need greater resolution, we have to reduce the step value (value of each bit). Let's say we make it 0.20 V instead of 1.0 V. Now, the recognized values will be

0000	0 V
0001	0.2 V
0010	0.4 V
0011	0.6 V
0100	0.8 V
0101	1.0 V
0110	1.2 V
0111	1.4 V
1000	1.6 V
1001	1.8 V
1010	2.0 V
1011	2.2 V
1100	2.4 V
1101	2.6 V
1110	2.8 V
1111	3.0 V

Notice that the range has been sharply reduced by the change in the step value. (Everything in life is a compromise, isn't it?) Instead of 0 to 15 V, we can now only cover 0 to 3 V with the same four bits.

There is only one way to increase the resolution without reducing the range (or vice versa). That is to increase the number of bits. If we use eight bits instead of four, we have 256 steps rather than the 16 we get with the four-bit system. If the step value is 1 V, the eight-bit range will be from 0 to 255 V. If a 0.20-V step value is used, the range will run from 0 to 51 V. Again, there is a price. Increasing the number of bits increases the cost and complexity of the A/D converter circuitry.

There is no law that says the range has to start at 0 V for a digital value of 0000. In the following examples, the step size is 0.15 V. A starts at 0 V, B starts at 2.50 V, and C starts at −1.00 V:

Digital value	A	B	C
0000	0.00	2.50	−1.00
0001	0.15	2.65	−0.85
0010	0.30	2.80	−0.70
0011	0.45	2.95	−0.55
0100	0.60	3.10	−0.40
0101	0.75	3.25	−0.25
0110	0.90	3.40	−0.10
0111	1.05	3.55	0.05
1000	1.20	3.70	0.20
1001	1.35	3.85	0.35
1010	1.50	4.00	0.50
1011	1.65	4.15	0.65
1100	1.80	4.30	0.80
1101	1.95	4.45	0.95
1110	2.10	4.60	1.10
1111	2.25	4.75	1.25

All three of these examples have the same step and range size, but the range is shifted.

One of the most common forms of A/D converter circuit is the single-slope circuit, shown in block diagram form in Fig. 13-6. The analog input signal is fed to an op amp that is wired as a comparator. The input signal is compared to the output of the D/A converter, which comes later in the circuit (a feedback loop).

The comparator puts out one bit, which is a one if and only if the input is greater than the output from the D/A converter. This bit is inverted and fed to an AND gate, along with the clock signal. There are four possible input combinations to this gate. Each input combination and its resulting output is as follows:

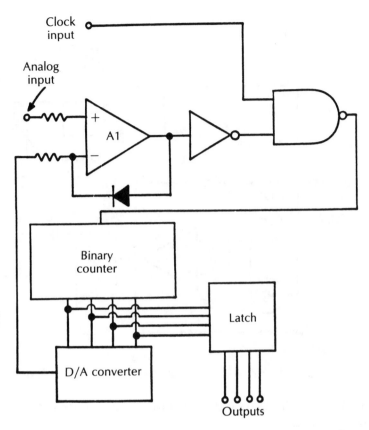

Fig. 13-6 *A relatively simple approach to A/D conversion is the single-slope converter.*

Comparator	Clock	Gate output
0	0	1
0	1	0
1	0	0
1	1	0

Notice that the gate output is a 1 if and only if both the clock and the inverted comparator output are at logic 0. (The direct converter output is a 1, meaning the analog input signal is greater than the D/A output.)

As long as the inverted comparator output is high, the clock signal is ignored. When the inverted comparator signal is low, the clock signal passes through the gate. The pulses are counted by the binary counter. The output of the counter is converted into

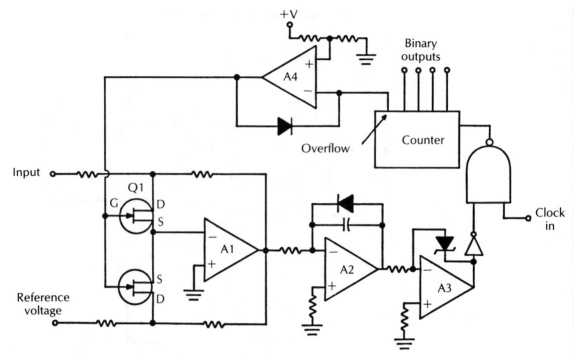

Fig. 13-7   *Better accuracy can be achieved with a dual-slope A/D converter.*

analog form by the D/A converter. This analog value is fed back to the comparator.

The counting continues until the D/A output exceeds the external input. The gate is now cut off, so no further clock pulses get through to be counted. The counter now holds a digital value that is proportional to the analog input voltage. This binary number is fed out to the computer input.

The single-slope A/D converter is functional, but its accuracy is not really the greatest. Better accuracy can be achieved with a dual-slope A/D converter, as illustrated in Fig. 13-7.

Obviously it takes some finite time for each sample value to be calculated. When faster conversions are required, a flash converter is used. This is simply a number of comparators connected in parallel. Each comparator contributes one weighted bit.

Index

Other Bestsellers of Related Interest

Electronic Power Control: Circuits, Devices, & Techniques
Irving M. Gottlieb
This guide focuses on the specific digital circuits used in electronic power applications. It presents state-of-the-art approaches to analysis, troubleshooting, and implementation of new solid-state devices. Gottlieb shows you how to adapt various power-control techniques to your individual needs. He uses descriptive analysis and real-world applications wherever possible, employing mathematical theory only when relevant.
0-07-023919-3 $27.95 Hard

**Upgrade Your IBM® Compatible and Save a Bundle,
2nd Edition**
Aubrey Pilgrim
Praise for the first edition . . .
"Every aspect is covered . . . liberally and clearly illustrated . . . invaluable."   **—PC Magazine**
Find valuable advice on adding the newest high-quality, low-cost hardware to our PC with this book. It offers informative how-to's for replacing motherboards with 80286, 80386, and 80486 boards; adding new floppy and hard disk drives; memory boards; and plugging in internal modems and VGA, fax, and network boards.
0-07-157784-X $19.95 Paper

The Alarm, Sensor, and Security Circuit Cookbook
Thomas Petruzzellis
Dozens of state-of-the-art circuit recipes for security systems, condition detectors, automobile alarms, portable and remote alarms, and more. An ideal circuit reference for technicians, hobbyists, and students. Covers computer interfacing techniques.
0-07-049707-9 $19.95 Paper
0-07-049706-0 $29.95 Hard

Electronic Projects to Control Your Home Environment
Delton T. Horn
More than 25 do-it-yourself projects for making affordable sensing and detection equipment. Includes projects like digital thermometers, humidity and temperature alerts, air ionizers, radiation and microwave monitors, and electronic pest repellents.
0-07-030417-3 $16.95 Paper
0-07-030416-5 $26.95 Hard

Build Your Own Intelligent, Mobile, SPACE Robot
Stephen James Montgomery
This groundbreaking guide outlines the process of designing and
building a SPACE (self-programming, autonomous,
computer-controlled, evolutionary/adaptive) mobile robot. Disk
contains self-programming software enabling robots to ''learn'' from
successes and failures.
0-07-042946-4   **$29.95 Paper**
0-07-042945-6   **$49.95 Hard**